T0328259

Environmental Kuznets Curve (EKC)

A Manual

Environmental Kuznets Curve (EKC)

A Manual

Edited by

BURCU ÖZCAN
Professor
Faculty of Economics and Administrative Sciences
Firat University
Elazig, Turkey

ILHAN ÖZTÜRK
Professor
Faculty of Economics and Administrative Sciences
Cag University
Mersin, Turkey

ELSEVIER

ACADEMIC PRESS
An imprint of Elsevier

Publisher: Candice Janco
Acquisition Editor: J. Scott Bentley
Editorial Project Manager: Ruby Smith
Production Project Manager: Kiruthika Govindaraju
Cover Designer: Alan Studholme

Academic Press is an imprint of Elsevier
125 London Wall, London EC2Y 5AS, United Kingdom
525 B Street, Suite 1650, San Diego, CA 92101, United States
50 Hampshire Street, 5th Floor, Cambridge,
MA 02139, United States
The Boulevard, Langford Lane, Kidlington, Oxford OX5 1GB,
United Kingdom

Working together
to grow libraries in
developing countries

www.elsevier.com • www.bookaid.org

List of Contributors

Buket Altinoz, PhD candidate
Department of Economics
Nevsehir Haci Bektas Veli University
Nevsehir, Turkey

Nicholas Apergis, PhD
Professor
Banking & Financial Management
University of Piraeus
Piraeus, Attica, Greece

Research Professor
Economics and Finance
University of Derby
Derby, United Kingdom

Alper Aslan, PhD
Professor
Department of Aviation Management
Erciyes University
Kayseri, Turkey

H. Hilal Bağlıtaş, PhD
Economics
FEAS
Erciyes University
Talas-Melikgazi, Kayseri, Turkey

Daniel Balsalobre, PhD
Faculty of Social Sciences
Department of Political Economy and Public Finance
Economic and Business Statistics and Economic Policy
University of Castilla-La Mancha
Cuenca, Spain

Mita Bhattacharya, PhD
Senior Lecturer
Department of Economics
Monash Business School
Caulfield, Australia

Faik Bilgili, PhD
Professor
Economics
FEAS
Erciyes University
Talas-Melikgazi
Kayseri, Turkey

Siew-Pong Cheah
Universiti Tunku Abdul Rahman
Kampar, Perak, Malaysia

Chee-Keong Choong
Universiti Tunku Abdul Rahman
Kampar, Perak, Malaysia

Eyup Dogan
Associate Professor
Department of Economics
Abdullah Gül University
Kayseri, Turkey

Pelin Gençoğlu, PhD
Economics
SSI
Erciyes University
Talas-Melikgazi, Kayseri, Turkey

Roula Inglesi-Lotz, PhD
Associate Professor
Department of Economics
University of Pretoria
Pretoria, South Africa

Abdul Jalil, PhD
Professor of Economics
Pakistan Institute of Development Economics
Quaid-i-azam University
Islamabad, Pakistan

Emrah Koçak, PhD
Associate Professor
Economics
Erciyes University
Kayseri, Turkey

Sevda Kuşkaya, PhD
Economics
SSI
Erciyes University
Talas-Melikgazi
Kayseri, Turkey

Lin-Sea Lau, PhD
Assistant Professor
Universiti Tunku Abdul Rahman
Kampar, Perak, Malaysia

Erhan Muğaloğlu, PhD
Economics
FBA
Abdullah Gül University
Kocasinan, Kayseri, Turkey

Cheong-Fatt Ng
Universiti Tunku Abdul Rahman
Kampar, Perak, Malaysia

Burcu Özcan
Professor
Faculty of Economics and Administrative Sciences
Firat University
Elazig, Turkey

Ilhan Öztürk
Professor
Cag University
Faculty of Economics and Administrative Sciences
Mersin, Turkey

Nasir Hamid Rao, MPhil Scholar
Pakistan Institute of Development Economics
Islamabad, Pakistan

Muhammad Shahbaz, PhD
Professor
Energy and Sustainable Development
Montpellier Business School
Montpellier, France
COMSATS University Islamabad
Lahore Campus
Pakistan

Avik Sinha, FPM
Department of General Management and
   Economics
Goa Institute of Management
India

Recep Ulucak, PhD
Associate Professor
Economics
Erciyes University
Kayseri, Turkey

Ali Gökhan Yücel, PhD
Erciyes University
Economics
Faculty of Economics and Administrative Sciences
Kayseri, Turkey

Preface

This book contains everything that students and researchers need to know about the environmental Kuznets curve, theory, and practice. Well-established academicians and researchers have united their knowledge expertise to provide this highly updated volume which constitutes an indispensable book belonging to the bookshelf of everyone who is fond and fascinated with the investigation of this verbose theory.

The environmental Kuznets curve is regarded as a statistical artifact that summarizes aspects of collective human behavior in terms of environmental pressure and sustainable development. Since the inauguration of this discussion with Kuznets who had hypothesized an inverted U-curve relationship between economic development and human inequality, numerous studies have been spawned, each employing different variables, methods, and research contexts, with the first one generated by Grossman and Kruger. After the 1990s the relationship has come to the foreground of environmental and resource economics (not least energy economics) postulating that the first stages of economic growth are characterized by increased exploitation of resources and the consequent contamination of the planet, but after a threshold level, environmental degradation will start declining through technological and institutional advancement.

Research in the field, among other things, estimates the crucial point where pressure to the environment starts lessening, after the human beings have consumed the essential material for its initial survival and progress. It is as if human beings have been granted a loan from Mother Nature and when the time comes, they have to pay it back. The environmental Kuznets curve theory can be very informative for the reversal of every possible externality applicable in economies.

This collective volume offers a tool for the economics and the econometrics of this emblematic theory. The economics part offers a comprehensive overview of the history and origin of this theory, its early and most recent studies, a separation of single and multiple country applications, policy landmarks, and the particular contribution of renewable energy to the solution of many energy resource exploitation problems. On the other hand, the econometrics part handles questions about data selection problems, time series, and panel series handling advice as well as the most cutting-edge econometric methods applicable in this field.

I am honored to provide a preface of this book and I am sure it will attract major readership. I hope you enjoy reading it as much as I have.

Angeliki N. Menegaki
Agricultural University of Athens, Athens, Greece
Open University of Cyprus, Nicosia, Cyprus

Introduction

We have been witnessing various environmental threats such as global warming, climate change, flood, drought, water scarcity, and so on since the industrialization age. In this respect, industrialization caused environmental pollution along with economic growth. Mechanization raised the production (output) levels of countries; however, it also damaged our habitat by increasing greenhouse gas emissions rates and ecological footprint level, by causing water pollution and scarcity, and so on. Those environmental problems could be evaluated as the side effects of economic development process. As will be discussed in the following subheading, scholars started interrogating the environmental effects of economic development process. Currently, two essential questions need to be answered: Does economic growth aggravate environmental destruction? Alternatively, does economic growth ameliorate this destruction with eco-friendly technologic improvements, that is, green-technology products?

Inspired by Simon Kuznets (1955), who explained the relationship between income inequality and economic growth with an inverted U-shaped curve, Grossman and Krueger (1991) analyzed the linkage between environmental pollution and economic growth. They asserted that in the early stages of development, environmental pollution first increases alongside the rising per capita income level. However, this trend turns reverse in the later stages of development process, that is, environmental pollution level starts going down while per capita income level goes up. Thereby, the traditional environmental Kuznets curve (EKC), which assumes an inverted U-shaped relationship between pollution and growth, has emerged. After that, several applied researchers questioned its validity for different countries, but they did not come to the same conclusion because of the differences in the economic development levels and the economic structures of countries.

Based on the interrelated natures of the economy and environment, this book deals with the specific topic entitled "the EKC" in the discipline of environmental economics. We aim to review both the theoretic and the empirical dimensions of the EKC subject by analyzing the studies in the existing literature. The book is a constructive synthesis of knowledge and experience from the experts who are internationally recognized for their work and contribution into the EKC literature. The book includes four main parts with several chapters.

Part 1: Introduction

Chapter 1. A Historical Perspective on Environmental Kuznets Curve

The chapter written by Prof. Burcu Ozcan and Prof. Ilhan Ozturk provides a historic perspective on global environmental concerns raising over time. The "limits to growth" report, the Brundtland Report, the sustainable development notion, and the EKC are the essential notions emphasized in the chapter. In particular, this chapter focuses on the emergence of the EKC concept and the reasons behind the inverted U-shaped curve.

Part 2: The Environmental Kuznets Curve Literature

Chapter 2. The Environmental Kuznets Curve: A Critical Review of Earlier Literature

In this chapter, Dr. Mita Bhattacharya critically analyses the earlier empirical studies between 1991 and 2006 within the EKC literature. The chapter highlights the major findings from the literature and indicates the limitations of these early studies. She also explains the key reasons behind the EKC, including income elasticity of environmental quality demand, scale, technologic, and composition effects.

Chapter 3. Recent Studies (Extending Basic Environmental Kuznets Curve Model by Adding More Variables)

Prof. Roula Inglesi-Lotz discusses the recent EKC studies in the literature that extend the bivariate EKC model with additional variables such as tourism, financial development, international trade, urbanization, and so on. The discussion evolves around the role that these factors play in the traditional bivariate EKC context and summarizes general findings of recent studies.

Chapter 4. Single-Country Versus Multiple-Country Studies

This chapter prepared by Prof. Alper Aslan, Assoc. Prof. Eyup Dogan, and Buket Altınoz provides an extensive literature review classified according to their county samples, that is, the single-country studies versus the multiple-country studies. They also explain the results and the methodologies of the studies along with their samples and time periods. The authors state that the EKC hypothesis has a long history and is still up to date.

Chapter 5. The Process of Sustainability: From Past to Present

In this chapter, Assoc. Prof Recep Ulucak, Dr. Ali Gokhan Yucel, and Assoc. Prof. Emrah Kocak review the international efforts on the process of sustainable development and environmental awareness by discussing alternative recommendations. The crucial world environmental summits and conferences, for example, the Stockholm Conference, Rio (Earth) Summit, Kyoto Protocol, World Summit on Sustainable Development, and Paris conference, are discussed. The authors also offer some alternative policy recommendations to solve environmental problems.

Chapter 6. Renewable Energy and Its Finance as a Solution to the Environmental Degradation

Prof. Nicholas Apergis suggests renewable energy as a solution to the environmental pollution problems. In this chapter, he explains some of the main topics regarding energy finance in the field of renewable energy and documents the "direction" of innovation that financial actors create.

Part 3: The Econometrics of the Environmental Kuznets Curve Studies

Chapter 7. Data Selection and Environmental Kuznets Curve Models - Environmental Kuznets Curve Models, Data Choice, Data Sources, Missing Data, Balanced and Unbalanced Panels

Dr. Avik Sinha, Prof. Muhammad Shahbaz, and Dr. Daniel Balsalobre prepared this chapter. The empirical part of the EKC studies was introduced in this chapter. They discuss crucial issues such as model selection, model validation, data standardization, variable selection and creation, and proxy for environmental degradation. While discussing these issues in brief, they also provide basic solutions to these problems both logically and empirically.

Chapter 8. Time-Series Analysis (Stationarity, Cointegration, and Causality)

This chapter written by Prof. Abdul Jalil and Nasir Hamid Rao focuses on the issues of time-series analysis such as spurious regression, stationarity, unit root with and without structural breaks, and cointegration with and without structural breaks in the framework of the EKC issue. They also provide some examples using these methods from the literature. They suggest that researchers who test the EKC hypothesis in the country cases should apply the newly developed time-series methodologies.

Chapter 9. Panel Data Analysis (Stationarity, Cointegration, and Causality)

In this chapter, Assist. Prof. Dr. Lau Lin-Sea, Ng Cheong-Fatt, Cheah Siew-Pong, and Prof. Choong Chee-Keong summarize both the conventional and contemporary panel data estimation techniques (e.g., unit root tests, cointegration tests, causality tests, and long-run estimators) used in the analysis of the EKC hypothesis. They state that previous studies on the relationship between pollution and economic growth have reached consistent or conflicting findings due to different estimation techniques. They argue that researchers have been able to provide justifications in explaining and predicting the EKC phenomenon by using the recently advanced panel data analyses such as Common Correlated Effect Mean Group (CCEMG) and Augmented Mean Group (AMG). They cite from the EKC studies using panel data models, as well.

Chapter 10. Most Up-to-Date Methodologic Approaches: Evidence from the Wavelet Coherence Approach.

Prof. Faik Bilgili, Dr. Erhan Mugaloglu, Dr. Sevda Kuskaya, H. Hilal Baglıtas, and Dr. Pelin Gencoglu prepared this chapter. They assert that mathematical calibrations, or time series and/or panel data estimations for EKC, in general, obtain the parameter estimations that do not change within the sample period. They suggest using wavelet coherence approach as a novel and a different methodology to examine the EKC by considering all possible shifts (structural breaks) in estimated parameters. They also provide an empirical study based on this methodology for the US economy.

Part 4: Conclusion.

Editors Prof. Ilhan Ozturk and Prof. Burcu Ozcan write this part. They provide general knowledge about the book and summarize its content.

Contents

CHAPTER 1

A Historical Perspective on Environmental Kuznets Curve

BURCU ÖZCAN • ILHAN ÖZTÜRK

A HISTORICAL PERSPECTIVE ON ENVIRONMENTAL KUZNETS CURVE

The close reciprocal linkage between environmental performance and economic development is of great importance and became a crucial subject of interest for too long. Regarding the economic impacts of environment, environmental quality may affect economic development through some channels. For instance, affecting the productivity level of labor, that is, the health conditions of workers, environment may have influences on the national human capital stock, which is the critical input for the development process. In this respect, Grossman and Krueger (1995, p. 355) argue that "our lives are affected by the air we breathe, the water we drink, the beauty we observe in nature, and the diversity of species with which we come into contact." This situation is an indication that environmental quality has an undisputed importance for human being. Besides, economic development also has some effects on the environment; however, there is no any unanimous idea whether economic development is a friend or a foe to the environmental welfare. Therefore, researchers and scholars have been trying to find answer to the question whether economic development can be used as a solution for the environmental problems.

Particularly, since the industrialization era, because of the rising public concerns over environmental issues and the deterioration of environmental quality worldwide, scientists started searching for the reasons and solutions of the environmental degradation problems. For instance, global warming is accepted as a by-product of excessive consumption of fossil fuels by humans after the industrial revolution in the mid-18th century (Uchiyama, 2016), and countries have been struggling to curb the greenhouse gas emission levels against global warming and climate change

problems. For the 21st century we are currently in, humanity has been witnessing some rising serious environmental disasters, such as flood, drought, melting of glaciers, global warming, and so on, which are the results of increasing pressure of human activities on the nature. Because of the environmental stress and pressure stemming from the human activities, the ecologic footprint level of each country increases as well. In other words, the ecologic footprint of human being, which is the total area necessary to produce the resources it consumes and to absorb the waste it generates based on prevailing technology (Bagliani, Bravo, & Dalmazzone, 2008), has surpassed the Earth's biocapacity, that is the amount of biologically productive land and water areas available within the borders of a given country (Moran, Wackernagel, Kitzes, Goldfinger, & Boutaud, 2008).

According to the Global Footprint Network (GFN)[1], the Earth's total biocapacity was 12.2 billion gha[2] (1.71 gha per person), whereas the humanity's ecologic footprint was 20.6 billion gha (2.87 gha per person) in 2013. Thus, the Living Planet Report prepared by the World Wide Fund for Nature (World Wide Fund for Nature (WWF), 2016) estimates that humanity currently needs the regenerative capacity of 1.6 Earths to provide the goods and services we consume each year. In this respect, a number of countries are moving from ecologic creditors to ecologic debtors, a phenomenon that ultimately manifests itself in global overshoot[3] (Wackernagel, Monfreda, Erb, Haberl, & Schulz, 2004). For instance, growing water

---

[1]See https://www.footprintnetwork.org/.

[2]Global hectares.

[3]Overshooting indicates the situation in which total demand for ecological goods and services exceeds the available supply for a given location (GFN, 2010).

Environmental Kuznets Curve (EKC), https://doi.org/10.1016/B978-0-12-816797-7.00001-1

shortages, desertification, erosion, reduced cropland productivity, overgrazing, deforestation, rapid extinction of species, collapse of fisheries, and global climate change are the consequences of ecologic overshoot.[4] In this respect, as stated by Galli et al. (2012), this century will be shaped by global overshoot, which will become increasingly evident in daily life in case that the trend of increasing human pressure do not reverse. Therefore, understanding the mechanism driving environmental problems is crucial to correctly assess the need and the usefulness of environmental policies because the carrying capacity of the earth may be surpassed sooner than we think (Urheim, 2009).

Even though we are living in a century accelerating environmental threats and catastrophes as explained above, the concerns about environmental degradation and natural resource depletion date back to the 18th century, the age of industrialization. For instance, in 1798, Thomas R. Malthus published a book—An Essay on the Principle of Population—in which he assumed that food supply grew arithmetically while population grew geometrically, and thus the world would be in misery because of the pressure of increasing population on the food supply. This situation was called the *Malthusian Population Trap Theory* where long-run human progress would be very dim (Ginevicius, Lapinskiene, & Peleckis, 2017). The theory emphasized the presence of a stage at which the food supply is inadequate for feeding the population given that population growth is ahead of agricultural growth. The Malthusian pessimism continued in the 20th century as well. For instance, in 1972, an international team of researchers from the Massachusetts Institute of Technology published the *"Limits to Growth"* report for the Club of Rome. They drew highly pessimistic scenario for the future of the world and underlined the reality that the rates of population growth, usage of resources, pollution increase, and material consumption would first grow exponentially but then collapse during the next century (Ekin, 1993; Tahvonen, 2000). The reason of the collapse is the limits that world economy will likely reach in terms of nonrenewable resources, agricultural production, and excessive pollution. Meadows, Meadows, Randers, and Behrens (1972, p. 23) emphasized the limits to growth as follows:

If the present growth trends in world population, industrialization, pollution, food production, and resource depletion continue unchanged, the limits to growth on this planet will be reached sometime within the next one hundred years.

They asserted that economic growth would be prevented as a result of finite environmental resource stock and that a steady-state economy with zero growth is a requirement to avoid the upcoming dramatic ecologic scenarios (Dinda, 2004). Therefore, they warned that patterns of production had to be changed from quantity to quality; otherwise, it would be impossible to satisfy the infinite needs of humanity (Ginevicius et al., 2017). Stressing the importance of the world's limited natural resource endowments, they advocated that world would be better off limiting its growth as opposed to continuing to reach for maximum growth in the long-run (Urheim, 2009).

In the course of time, there happened some serious events such as the oil crises in the 1970s confirming the ideas supported by the Club of Rome. Oil prices made peak because of oil shortage and resulted in the world energy crisis. After the oil crises, there were only few who questioned the view that the world was entering a future of rising scarcity of energy and natural resources (Tahvonen, 2000). However, though this pessimistic atmosphere, during the 1970s, some empirical studies supported that the ratio of consumption of some metals to income was declining in the developed economies, which conflicted with the predictions introduced in the Limits to Growth Report (Malenbaum, 1978). Owing to positive or negative environmental developments, The United Nations Conference on the Human Environment, also known as the Stockholm Conference, was held in Stockholm, Sweden from June 5 to 16, 1972. It was the first major conference of the United Nations on the international environmental issues and marked a turning point in the development of international environmental awareness.[5]

In the 1980s, humanity witnessed much progress and many improvements in the science and technology fields that eliminate the pessimistic considerations about the environment and natural resource endowments. For instance, economics discipline introduced a new economic growth theory, named as "the endogenous growth theory," which assumes that technologic development is a continuous progress stemming from innovations made in firms and could be supported by the governments research and development investments and expenditures (see, inter alia, Grossman & Helpman, 1991; Lucas, 1988; Romer, 1986). Thus, technologic development that produces eco-friendly technologic devices was accepted as a remedy for the

---

[4]See https://www.wwf.at/de/view/files/download/showDownload/?tool=12&feld=download&sprach_connect=1814.

[5]See https://sustainabledevelopment.un.org/milestones/human environment.

environmental problems and natural resource depletion. Technologic developments created two conflicting views: On one side, scientists stated that consumption could be sustained through substitution and technical progress even if production relies on finite natural resource (Klaassen & Opschoor, 1991) and on the other side, environmental economists argued that substitution possibilities for scarce resources are restricted by physical laws even in the case of continuous technologic change (Tahvonen, 2000).

As an output of the aforementioned environmental discussions, the *sustainable development* concept became an essential doctrine of the economic and political agenda in the 1980s. For instance, in 1987, the World Commission on Environment and Development (WCED) published a report, namely "Our Common Future," which is also known as the Brundtland Report. Herein, the concept of "sustainable development" replaced the concept of "limits to growth," and the emphasis was placed on complementarity between growth and environment (Ekin, 1993). The WCED (1987, p. 41) defined sustainable development as the "development that meets the needs of the present without compromising the ability of future generations to meet their own needs." In this respect, the *intergenerational equity* notion, indicating that we inherit the Earth from previous generations and have a duty to pass it on in good condition to future generations[6], has gained importance on the way of sustainable development. The concept of intergenerational equity means that humanity needs to promise to future generations the same level of economic prosperity and environmental conditions that the present generation has (Uchiyama, 2016). The sustainable development concept aims at maintaining economic advancement while protecting the long-term value of the environment (Emas, 2015). As an essential part of the sustainable development argument, the Brundtland Report promulgates the idea that economic growth is necessary for the permanent environmental quality.

Moreover, the harmony between population size and economic growth, with the changing productive potential of the ecosystem, is suggested as a key to follow a sustainable development path. Sustainable development currently serves as the basis of environmental policies given that it assumes a peaceful coexistence between environment and economic development, rather than a trade-off relationship

(Uchiyama, 2016). The long-term stability between economy and environment is accepted as the main goal of sustainable development. Additionally, the integration and acknowledgment of economic, environmental, and social concerns throughout the decision-making process is accepted as a requirement to acquire the sustainable development target (Emas, 2015). The notions that development does not require environmental destruction and that poverty reduction is essential for environmental protection are the fundamental doctrines of sustainable development (Stern, 2015). The WCED (1987, p. 16) emphasizes the importance of poverty reduction for the future of humanity as follows:

> *Poverty is not only an evil in itself, but sustainable development requires meeting the basic needs of all and extending to all the opportunity to fulfill their aspirations for a better life. A world in which poverty is endemic will always be prone to ecological and other catastrophes.*

In the historical perspective, another milestone was the "World Development Report (1992)" of the World Bank, prepared by a team consisting of several scholars. The World Development Report (1992) suggested the integration of environmental considerations into development strategy as a fundamental policy tool because underestimation of the value of environment for a long time has damaged human health, reduced productivity, and undermined the future development prospects. Besides, the Report aimed at providing compromise between development and environment by offering equitable income growth as a solution for environmental threats resulted from the growth process. As defined in the following quotation, a twin strategy is proposed to achieve both the economic growth and the environmental conservation targets:

> *Some problems are associated with the lack of economic development; inadequate sanitation and clean water, indoor air pollution from biomass burning, and many types of land degradation in developing countries have poverty as their root cause. Here the challenge is to accelerate equitable income growth and promote access to the necessary resources and technologies.*
>
> **WDR, 1992, P. 7**

Moreover, the Report presumed environmental protection as an essential part of economic development (Stern, 2004b) while the absence of development is accepted as a reason of environmental destruction (Ginevicius et al., 2017) because resources necessary for investment would be inadequate without development. In this regard, the Report underlined the importance of development to achieve a high

---

[6]See https://www.uow.edu.au/~sharonb/STS300/equity/meaning/integen.html.

level of environmental quality with the following sentences:

The view that greater economic activity inevitably hurts the environment is based on static assumptions about technology, tastes, and environmental investments (WDR, p.38). …. As income rises, the demand for improvements in environmental quality will increase, as will the resources available for investment

WDR, 1992, P. 39

Furthermore, WDR Report popularized the environmental Kuznets curve (EKC) concept as well, and in the early 1990s with the Grossman and Krueger's (1991) path-breaking study about the potential impacts of NAFTA[7] (Stern, 2004a, 2004b; Urheim, 2009), the EKC has become a popular research field among analysts modeling the nexus of ambient pollution concentrations and economic growth (Stern, 2015). In a cross-section of urban areas located in 42 countries, Grossman and Krueger (1991) studied the relationship between air quality and economic growth in the context of liberalization of trade (NAFTA) between the United States and Mexico. They supported that trade liberalization affects environment by expanding the scale of economic activity as well as modifying its composition and bringing about a change in the production methods. They confirmed the EKC only for two air pollutants, that is, sulfur dioxide and smoke. In another study, Grossman and Krueger (1995) continued to investigate the environmental effects of economic growth and asserted that development creates a number of structural changes in the production process through which societies may find ways to conserve their finite natural resource stocks. They believed that harmful effects of economic growth on the environment could be compensated by the technologic improvements and innovations. For the most pollutants, they found that environmental pollution increases at the initial stage of economic growth, but a final phase of improvement could be observed in the course of the economic growth. The increased demand and supply for environmental protection at higher income levels is accepted as the essential reason of the improvement.

If we look at the origin of the EKC, we can see that it was essentially inspired by Kuznets (1955). In his original study, Kuznets (1955) asserted that income inequality first increases, then reaches a turning point, and declines thereafter during the evolution of an economy from a low-income country to a high-income country (Diet, Rosa & York, 2012). Therefore, an inverted U-shaped relationship is assumed between income inequality and economic growth. In other words, the distribution of income is unequal in the early stages of economic growth, whereas it becomes more equal in the later stages of growth (Kuznets, 1955). Based on the logic of the original Kuznets curve, the EKC hypothesis assumes that further economic growth is likely to improve environmental performance after countries surpass an adequate level of economic growth. As national economies develops, the pressure they created on the nature increases at first but eventually reaches a turning point beyond which further growth decreases environmental pressure (Dietz et al., 2012). This implies that environmental pollution is an inverted U-shaped function of income per capita. As Dinda (2004) noted, throughout out an environmental Kuznets curve, we can understand how a technically specified measurement of environmental quality changes as the fortunes of a country. In this regard, the EKC is a theoretical proposition explaining why nations may change over time in creating environmental pressure and why nations differ from each other in the environmental pressure they generate (Dietz et al., 2012).

Since the early period advocates of the EKC (see, inter alia, Grossman & Krueger, 1991, 1995; Panayotou, 1993; Selden & Song, 1994; Shafik & Bandyopadhyay, 1992), the relationship between economic growth and environmental pollution has been assumed shifting throughout the development paths of economies. The effect of economic growth on the nature may change from negative to positive as long as countries achieve higher income levels at which people demand and afford more efficient infrastructure and cleaner habitats (Panayotou, 2003). If the EKC hypothesis is true, economic growth is a solution for the environmental pollution problems, rather than being a threat to the environment. However, although advocates of the EKC and the sustainable development goal are embraced by all countries, there is still doubt about the coexistence of economic growth and environmental improvements. For instance, Ginevicius et al. (2017) supported the idea that sustainable development is an oxymoron because humanity has been facing severe environmental problems and the natural habitat has been destroyed beyond repair since last four decades.

In respect of the stages of development, there are three essential stages, namely the preindustrial phase, the industrial phase, and the postindustrial phase, at which economic growth affects the environmental quality in different ways (see Fig. 1.1). In the early stage of development, that is, the period of preindustrial

---

[7]North American Free Trade Agreement.

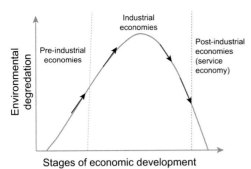

FIG. 1.1 The environmental Kuznets curve: a development-environment relationship. (Credit: Panayotou, T. (2003). Economic growth and the environment. Harvard University and Cyprus International Institute of Management. Available from: https://www.unece.org/fileadmin/DAM/ead/sem/sem2003/papers/panayotou.pdf.)

economies, environmental degradation level increases, but beyond a critical level of income per capita that reaches the turning point in industrial economies, which is the second stage of development, the trend starts reversing (Ginevicius et al., 2017). In the preindustrial and industrial phases, as agriculture and resource extraction intensify and industrialization takes off, the resource depletion and the waste generation accelerate (Dinda, 2004). As the economy is industrialized and starts producing manufacturing goods, pollution will likely increase monotonically with output (Urheim, 2009). In the early stage of industrialization, that is, at low-income levels, pollution would grow rapidly because people are struggling to satisfy their basic needs such as food and shelter. Additionally, as stated by Dasgupta, Laplante, Wang, and Wheeler (2002), in the early stage of industrialization, environmental regulations are weak and communities are not rich enough to pay for environmental pollution abatement strategies. Therefore, at low-income levels, little or no funds are allocated to environmental protection because poverty is pervasive, tax collection is ineffective, and environmental awareness level is low (Panayotou, 2003). In this respect, increased pollution is regarded as an acceptable side effect of economic growth in this phase (Arrow et al., 1995). In this scenario, for developing countries, there is a trade-off between economic growth and environmental quality because slowing economic growth in the interests of protecting environment does not have a priority on their agendas (Urheim, 2009). However, in the later stages of industrialization, along with rising income level, people attach more importance to environmental amenities and

the regulatory institutions become more effective, which results in less environmental pollution (Dietz et al., 2012; Hervieux & Darne, 2015). The consumers with higher income levels are willing both to spend more for green products and to create more pressure for the environmental protection and the regulations (Dinda, 2004). As stated by Grossman and Krueger (1995), wealthy countries have relatively more stringent environmental standards (or enforcement of laws) than the middle-income and poorer countries have. In this context, environmental quality is assumed to be a luxury good because it does not constitute a significant part of the consumer's budget until a high level of income is achieved (Wang, 2017). This situation indicates that income elasticity of environmental quality demand is in excess of unity.

Last, at the final phase of the development, named as the postindustrial economies, economic growth leads to environmental improvement because of increased demand for environmental quality and structural changes towards information-based industries and more efficient technologies (Panayotou, 2003). In other words, there is a shift from the more resource-intensive extractive and heavy industrial sectors towards services and lighter manufacturing, which reduce emissions per unit of output (Stern, 2004a).

In a harmonious way with the development phases, beginning with Grossman and Krueger (1991) and Panayotou (1993), the EKC literature has suggested three different channels through which economic growth may affect environmental welfare. First, the *scale effects* indicate that growth in the scale of the economy would result in a proportional growth in the environmental pollution and destruction (Panayotou, 1993; Stern, 2004a, 2004b). More input and natural resources are used to increase output as a result of expanding production scale. However, more output also contributes to environmental degradation because, as a by-product, it creates more waste and emissions (Dinda, 2004). In this respect, Grossman and Krueger (1991) asserted that total amount of pollution increases as a result of expansive economic activity stemming from trade and investment liberalization if the nature of that activity remains unchanged. Therefore, the scale effects indicate that economic development and environmental quality are the two conflicting goals. Second, the *composition effects* have an inverted U-shaped relationship with income level as they indicate sectoral structural changes. As such, at low-income levels, the dominant shift is from the agriculture to industry with an increase of pollution intensity, whereas at high-income levels, the dominant shift is from the industry to services with a decrease in pollution intensity (Akbostanci, Asik, & Tunc, 2009). Therefore, environmental pollution first increases and then decreases with rising income level as long as the share of industry first goes up and then goes down (Panayotou, 2003). Third, the *technique effects*, appeared in the postindustrial economies, describe the change in resource and emission intensity of production because of technologic modernization (Bagliani et al., 2008). Further economic growth causes technologic progress through which dirty and obsolete technologies are replaced by the upgraded and cleaner technologies that improve environmental quality (Dinda, 2004). Thus, environmental degradation starts lessening based on the development of services and knowledge-based technology-intensive industries.

Overall, based on the abovementioned ideas and views, economic development appears as a double-edged sword because it has pros and cons for environmental performance. On one hand, development provides more technologically developed and eco-friendly devices; on the other hand, it sacrifices environment for further growth target, for example, through the installation of nuclear power plant. However, there is currently an increasing public awareness about environmental problems resulted from human activities. Particularly, the inhabitants of the developed Western countries are well aware of the upcoming environmental threats and take precautions against them. For instance, as emphasized by Urheim (2009), consumers currently know that an airplane emits more CO_2 than a train does traveling the same distance, and that the environment will be better off if public transportation is preferred to private cars. Therefore, we believe that there is still hope and light at the end of the tunnel to solve and lessen the environmental problems stemming from our own activities.

CONCLUSION

In general, since the beginning of the humanity, but in particular, since the industrial revolution in the 18th century, we have been witnessing a number of serious environmental degradation problems such as acid rain, global warming, water scarcity and pollution, melting glaciers, landslides, increasing frequency of droughts and floods, and so on. These problems are actually the results of human beings' own activities. We have polluted our habitat while trying to achieve

a more civilized and developed world, and therefore for a long time, we have been looking for solutions to ease our burden on the environment. In this chapter, focusing on the development and pollution nexus, we tried to provide a historical perspective, starting from the 18th century and continuing until the 21st century, on environmental challenges and struggles undertaken by countries. The debates and views expressed in this chapter interrogate the idea whether economic development is a foe or a friend of the environment. In this regard, based on the EKC hypothesis that is proposed by Grossman and Krueger (1991), researchers have focused on examination of close linkages between economic development and environmental pollution. EKC assumes an inverted U-shaped relationship between development process and pollution, that is, environmental pollution increases with the rising per capita income at the first phases of the development process; however, the pollution level begin to fall beyond a threshold income level reached at the later stages of development. As well as offering a historical perspective on environmental problems, this chapter also clarifies the reasons behind the inverted U-shaped EKC by explaining the scale, composition, and technique effects.

REFERENCES

Akbostanci, E., Asik, S., & Tunc, G. I. (2009). The relationship between income and environment in Turkey: Is there an environmental Kuznets curve. *Energy Policy, 37*(3), 861−867.

Arrow, K., Bolin, B., Costanza, R., Dasgupta, P., Folke, C., Holling, C. S., et al. (1995). Economic growth, carrying capacity and the environment. *Science, 268*(5210), 20−521.

Bagliani, M., Bravo, G., & Dalmazzone, S. (2008). A consumption-based approach to environmental Kuznets curves using the ecological footprint indicator. *Ecological Economics, 65*, 650−661.

Dasgupta, S., Laplante, B., Wang, H., & Wheeler, D. (2002). Confronting the environmental Kuznets curve. *The Journal of Economic Perspectives, 16*(1), 147−168.

Dietz, T., Rosa, E. A., & York, R. (2012). Environmentally efficient well-being: Is there a Kuznets curve? *Applied Geography, 32*, 1−28.

Dinda, S. (2004). Environmental Kuznets curve hypothesis: A survey. *Ecological Economics, 49*, 431−455.

Ekins, P. (1993). 'Limits to growth' and 'sustainable development': Grappling with ecological realities. *Ecological Economics, 8*, 269−288.

Emas, R. (2015). *The concept of sustainable development: Definition and defining principles.* Available from: https://www.academia.edu/31158981/The_Concept_of_Sustainable_Development_Definition_and_Defining_Principles.

Galli, A., Wiedmann, T., Ercin, E., Knoblauch, D., Ewing, B., & Giljum, S. (2012). Integrating Ecological, Carbon and Water Footprint into a "Footprint Family" of Indicators: Definition and Role in Tracking Human Pressure on the Planet. *Ecological Indicators, 16*, 100−112. https://doi.org/10.1016/j.ecolind.2011.06.017.

Ginevicius, R., Lapinskiene, G., & Peleckis, K. (2017). The evolution of the environmental Kuznets curve concept: The review of the research. *Panoeconomicus, 64*(1), 93−112.

Grossman, G. M., & Helpman, E. (1991). Trade, knowledge spillovers, and growth. *Economic Review, 35*, 517−526.

Grossman, G. M., & Krueger, A. B. (1991). *Environmental impacts of a North American free trade agreement*. NBER Working Papers No. 3914.

Grossman, G. M., & Krueger, A. B. (1995). Economic growth and the environment. *The Quarterly Journal of Economics, 110*(2), 353−377.

Hervieux, M. S., & Darne, O. (2015). Environmental Kuznets curve and ecological footprint: A time series analysis. *Economics Bulletin, 35*(1), 814−826.

Klaassen, G. A. J., & Opschoor, J. B. (1991). Economics of sustainability or the sustainability of economics: Different paradigms. *Ecological Economics, 4*(2), 93−115.

Kuznets, S. (1955). Economic growth and income inequality. *The American Economic Review, 45*, 1−28.

Lorente, D. B., & Herranz, A. A. (2016). Economic growth and energy regulation in the environmental Kuznets curve. *Environmental Science and Pollution Research, 23*, 16478−16494.

Lucas, R. E. (1988). On the mechanics of economic development. *Journal of Monetary Economics, 22*, 3−42.

Malenbaum, W. (1978). *World demand for raw materials in 1985 and 2000*. New York: McGraw-Hill.

Malthus, T. (1798). *An essay on the principle of population as it affects the future improvement of society*. London: Ward Lock.

Meadows, D. H., Meadows, D. L., Randers, J., & Behrens, W. W. (1972). *The limits to growth: A report for the club of Rome's project on the predicament of mankind* (5th ed.). New York: Universe Books.

Moran, D. D., Wackernagel, M., Kitzes, J. A., Goldfinger, S. H., & Boutaud, A. (2008). Measuring sustainable development — nation by nation. *Ecological Economics, 64*, 470−474.

Panayotou, T. (1993). *Empirical tests and policy analysis of environmental degradation at different stages of economic development*. Geneva: International Labor Office. World Employment Programme Research, Working Paper No: 238.

Panayotou, T. (2003). *Economic growth and the environment*. Harvard University and Cyprus International Institute of Management. Available from: https://www.unece.org/fileadmin/DAM/ead/sem/sem2003/papers/panayotou.pdf.

Romer, P. M. (1986). Increasing returns and long-run growth. *Journal of Political Economy, 94*(5), 1002−1037.

Selden, T., & Song, D. (1994). Environmental quality and development: Is there a Kuznets curve for air pollution emissions. *Journal of Environmental Economics and Management, 27*, 147−162.

Shafik, N., & Bandyopadhyay, S. (1992). *Economic growth and environmental quality time-series and cross-country evidence*. Washington: The World Bank. Background Paper for the World Development Report 1992.

Stern, D. I. (2004a). The rise and fall of the environmental Kuznets curve. *World Development, 32*(8), 1419−1439.

Stern, D. I. (2004b). Environmental Kuznets curve. *Encyclopedia of Energy, 2*, 517−525.

Stern, D. I. (2015). *The environmental Kuznets curve after 25 years*. Australian National University. Centre for Climate Economics & Policy Working Paper 1514.

Tahvonen, O. (2000). *Economic sustainability and scarcity of natural resources: A brief historical review*. Washington: Resources for the Future. Discussion Papers.

Uchiyama, K. (2016). *Environmental Kuznets curve hypothesis and carbon dioxide emissions*. Springer. Springer Briefs in Economics Development Bank of Japan Research Series.

Urheim, M. (2009). *Economic growth and the environment: Friends or foes? The environmental Kuznets curve for CO_2 emissions*. University of Oslo. Master thesis in Environmental and Development Economics.

Wackernagel, M., Monfreda, C., Erb, K. H., Haberl, H., & Schulz, N. B. (2004). Ecological footprint time series of Austria, the Philippines, and South Korea for 1961−1999: Comparing the conventional approach to an 'actual land area' approach. *Land Use Policy, 21*, 261−269.

Wang, M. N. (2017). *Investigating the environmental Kuznets curve of consumption for developing and developed countries: A study of Albania and Sweden*. Bachelor's dissertation. Aalto University.

World Commission on Environment and Development (WCED). (1987). *Our common future*. Oxford: Oxford University Press.

World Development Report. (1992). *Development and the environment*. New York: Oxford University Press.

World Wide Fund for Nature (WWF). (2016). *Living planet report 2016: Risk and resilience in a new era*. Gland, Switzerland: WWF International.

CHAPTER 2

The Environmental Kuznets Curve: A Critical Review of Earlier Literature

MITA BHATTACHARYA, PHD

INTRODUCTION

Improving environmental quality is one of the most challenging tasks in our time. Over the past five decades, environmental quality has been associated with economic growth. Since the early 1970s, environmentalists and economists have raised awareness of sustainable environments (Meadows, Meadows, Randers, & Behrens, 1972) and emphasized how limited environmental resources would have a dampening effect on economic growth. These researchers urged a zero-growth or steady-state economy to avoid ecologic disasters. On the other hand, the role of technologic progress and the substitutability of natural and man-made capital has been emphasized by Beckerman (1992) for reducing the dependence on natural resources, indicating an everlasting growth path.

More recently, in the 2015 Paris climate change agreement, long-run environmental sustainability was established as the key objective for many countries. In particular, reduction of carbon emissions is high on the agenda, along with other pollutants. Pioneering research by William Nordhaus and Paul Romer was recognized by the 2018 Nobel Prize in Economics for integrating the important roles of climate change and technology into sustainable economic growth.[1] This year's Laureates' research led us to think in considerably more detail about achieving sustainable global economic growth.

In empirical energy economics research, the environmental Kuznets curve (EKC) hypothesis is perhaps the mostly tested phenomenon, offering the simplest framework linking energy use (and related negative spillover effects) with economic growth. A report from the World Bank (1992) first introduced this idea. Researchers used a simple empirical analysis, regressing data on ambient air and water quality in cities across the globe with a polynomial of GDP per capita and other city and country characteristics as controls. Plotting the fitted values of pollution levels as a function of GDP per capita, the researchers established inverse U-shaped EKC for many of the plots. The peaks of these predicted pollution-income paths vary across pollutants; however, "in most cases they come before a country reaches a per capita income of $8000" (in terms of 1985 dollars) (Grossman & Krueger, 1995, p. 353).

The literature is borrowed from the original Kuznets curve (https://www.ncbi.nlm.nih.gov/pmc/articles/PMC4158735/; Kuznets, 1955, 1998), reflecting the relationship between income and inequality. The EKC hypothesis describes the relationship between various indicators of environmental pollutants and income per capita. The common thread running behind this hypothesis is to examine a quadratic relationship between some indicators of environmental degradation and per capita income in testing the inverted U-shaped curve.[2] Looking from policy perspective, we question: can economic growth be part of the solution for rather than a cause of environmental problems?

---

[1]https://www.nobelprize.org/prizes/economic-sciences/2018/prize-announcement/.

[2]SO_2, per capita emissions of CO_2, annual deforestation, and urban sanitation are considered as various indicators of environmental degradation.

Environmental Kuznets Curve (EKC) https://doi.org/10.1016/B978-0-12-816797-7.00002-3

Although the literature on the EKC hypothesis has burgeoned in the past three decades, no concrete conclusions have been drawn. There is no consensus regarding the existence of an EKC, as shapes varied across the measures of environmental degradation and econometric techniques used in the studies across countries. Pearson (1994), Stern, Common, and Barbier (1996) and Barbier (1997) surveyed early studies on the literature, whereas Stern (2004) and Dinda (2004) cover both theoretic and empirical aspects in a critical review. Given that the focus of this chapter is on earlier studies analyzing the EKC hypothesis, the chapter covers studies between 1990 and 2006.

In Section 2.1.2, we briefly review the literature covering studies from various countries between 1990 and 2006. Section 2.1.3 describes on econometric models, highlighting criticisms of these models. Section 2.1.4 summarizes major observations in the context of policy relevance along with concluding remarks.

A REVIEW OF EARLY LITERATURE: 1990–2006

The literature on the EKC hypothesis began in the early 1990s and has evolved since the first empirical study by Grossman and Krueger (1991). Using the global environment monitoring system (GEMS) database, the researchers identified the potential environmental effects of the North American Free Trade Agreement (NAFTA). The turning points for SO_2 and dark matter were at around $4000 to 5000 while the concentration of suspended particles had a declining trend even with low-income levels. At income levels above $10,000–15,000, the estimates reflect an increasing trend for all three pollutants.

The findings from Shafik and Bandyopadhyay's (1992) research were used in the 1992 World Development Report (IBRD, 1992).[3] The research estimated the EKC models for ten different environmental indicators using log-linear, log quadratic, and a logarithmic cubic polynomial in GDP per capita, with time-trend and trade, politics, and infrastructure-related variables. In each case, the dependent variable was not transformed. Researchers established that lack of clean water and urban sanitation decline uniformly with increasing level of income over time. Both measures of deforestation were found to be insignificantly related with income: observed increase in income worsens the quality of rivers. The two air pollutants confirm the EKC

hypothesis. The turning points for both pollutants were between $3000 and $4000. Finally, both carbon emissions per capita and municipal waste increased unambiguously with rising income.

Following Grossman (1995), during the first stage of development, growth exhibits a *scale effect* on the environment. This effect is detrimental on natural resources and worsens the environmental quality. Possible explanations behind the downward sloping section of the EKC include that it is due to a *composition effect* of the economy. With the progress of economic development, economies face structural changes from clean agrarian to polluting industrial, then the clean service sector; more resources are invested to improve environmental quality. Panayotou (1993) emphasizes "at higher levels of development, structural change toward information-intensive industries and services, coupled with increased environmental awareness, enforcement of environmental regulations, better technology and higher environmental expenditures, result in leveling off and gradual decline of environmental degradation" (Panayotou, 1993, p. 1).

Finally, developed countries invest more into research and development, replacing obsolete equipment and technology with cleaner ones. This *technique effect* has a positive spillover into economic growth and development. Within the stages of economic development, any negative influence from the scale effects on the environment is outweighed by the positive impact of composition and technique effects. This result implies that income elasticity of environmental quality demands increases over time when a country passes through developmental stages; this effect has been emphasized by researchers (Beckerman, 1992; Carson, Jeon, & McCubbin, 1997; McConnell, 1997). With an increase in living standards, greener environmental resources and technologies are used in the production process. Such effects lead to the decreasing path of the EKC (Grossman, 1995). Table 2.1 summarizes the major relevant empirical studies between 1991 and 2006.[4]

Most early studies used panel estimation; the shape of the EKC remains ambiguous with different types of pollutants. Additionally, there is no agreement within the literature on the income level where the turning point starts. Cole, Rayner, and Bates (1997) pointed out that the alleged turning point has a very large standard error for most studies with a peak in the CO_2 curve. Specific features of any particular country may

---

[3]The International Bank for Reconstruction and Development (IBRD).

[4]We include the studies published during this period and are only based on panel estimations. Here our focus was to include studies with different pollutants not only on CO_2 emissions.

TABLE 2.1 The Environmental Kuznets Curve: An Overview of Empirical Literature Between 1991 and 2006.

Author(s)	Environmental Indicator(s)	Major Variables	Study Period(s)	Econometric Techniques	Major Findings
Grossman and Krueger (1991)	SO_2, smoke, mass of suspended particles	GDP per capita,	1988, NAFTA countries	Random effect model	For SO_2, smoke, inverted U-shaped is followed
Shafik and Bandyopadhyay (1992)	Deforestations, per capita CO_2, water	GDP per capita, a time trend	149 countries, 1961–86	Panel data, fixed effect model	CO_2, monotonically increasing
Panayotou (1993)	SO_2, NO_x, solid particulate matter (SPM) and deforestation	GNP per capita, population	Mid-1980s, 50 developing and developed countries	OLS estimation	Inverted U-shaped
Holtz-Eakin and Selden (1995)	Per capita CO_2	GDP per capita	130 countries, 1951–86	Panel data, fixed effect model	Quadratic inverted U-shaped, cubic N-normal
De Bruyn et al. (1998)	Sulfur emission	GDP per capita	1990 data to forecast target for 2000, LRTAP countries and the United States[5]	OLS	Greater reduction for high income countries
Suri and Chapman (1998)	Emissions from commercial manufacturing	GDP per capita, export and import shares of manufacturing	33 countries, 1971–91	Panel data, FGLS	Manufactured exports by industrialized countries has emission enhancing effect while manufactured imports by industrialized countries have contributed to the downward slope
Cole (2004)	Per capita CO_2; 9 more pollutants; water pollutants	GDP per capita, trade intensity, share of pollution intensive exports and imports on trade, share of manufacturing	21 OECD countries, 1980–87	Panel data; generalized least squares, random and fixed effect models	Inverted U-shaped for CO_2, for other pollutants both inverted U-shaped and N-shaped.
Martinez-Zarzoso and Bengochea-Morancho (2004)	CO_2 emissions	GDP per capita	22 OECD countries, 1975–98	Pooled mean squares	Both inverted U-shaped and N-shaped relationship for CO_2 emissions
Dinda and Coondoo (2006)	Per capita CO_2	GDP per capita, a time trend	88 countries, 1960–90	Panel data, cointegration, error correction test	Bidirectional relationship
Richmond and Kaufmann (2006)	Carbon emissions	GDP per capita	Panel of OECD and non-OECD countries	Pooled OLS and panel cointegration	Limited support of inverted U-shape for OECD countries, positive relationship between income-carbon emissions for non-OECD countries

[5] Long-range transboundary air pollution (LRTAP) countries.

not be generalized in cross-country studies. Roberts and Grimes (1997) and Vincent (1997) emphasized that the inverted U-shaped EKC across countries is a statistic artifact, not capturing various developmental stages for different countries or regions within a country. Galeotti, Lanza, and Pauli (2006) portray the early literature and its emphasis on functional form, estimation, and the limitation of data used in these studies. Many authors rightly warn about the nonstructural nature of the relationship, based mostly on reduced-form models.

ECONOMETRIC MODEL AND RELATED CRITICISMS

The earliest literature on the EKC models was with simple quadratic functions of levels of income. Following the original Grossman and Krueger (1991) research, some studies used a cubic EKC in levels and established an N-shaped EKC. This result may simply be a polynomial approximation to a logarithmic curve. The standard reduced-form EKC model can be stated as:

$$\ln(CO_2/P)_{it} = \alpha_i + \gamma_t + \beta_1\ln(GDP/P)_{it} + \beta_2\ln(GDP/P)_{it}^2 + \beta_3\ln(GDP/P)_{it}^3 + \varepsilon_{it}$$

where CO_2 is an indicator of environmental pollution, P is a population, and ln indicates natural logarithms. The first two terms on the RHS represent intercept parameters, which vary across countries (or regions), i, and years, t. α_i implies that the level of CO_2 emissions per capita may vary across countries at any income level, the income elasticity is the same for all countries at a given income level. γ_t, the time-specific intercept, counts time-varying omitted variables and common stochastic shocks to all countries. We assume, the EKC function is linear, quadratic and cubic functions in the simplest version of the model. Results imply, for an inverted U-shaped curve, $\beta_1 > 0$; $\beta_2 < 0$; and for an N-shaped EKC curve, $\beta_1 > 0$; $\beta_2 < 0$; $\beta_3 > 0$. The income level where "turning point" occurs, where emissions are at a maximum, is

$$\mu = -\beta_1/(2\beta_2)$$

In addition, a few papers have used a nonparametric approach using Kernel regressions (Azomahu & Van Phu, 2001; Taskin & Zaim, 2000) or a flexible parametric approach (Dijkgraaf & Vollebergh, 2001; Schmalensee, Stoker, & Judson, 1998).

This conventional approach in the early studies of the EKC remained as a "black box," where, for any particular country, income level was used as a catch-all surrogate variable incorporating all the changes

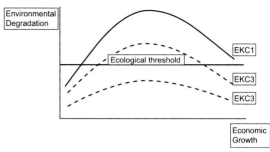

FIG. 2.1 EKC1 traces the path of environmental degradation when technological progress, policies, regulations, and other barriers in improving environment are not internalized into the growth process. EKC2 and EKC3 represent the path when all these factors are endogenous into the growth process. (Adapted from Panayotou, T. (1993). Empirical tests and policy analysis of environmental degradation at different stages of economic development (No. 992927783402676). International Labour Organization.)

with economic development. Fig. 2.1 shows a simple plot of the EKC indifferent scenarios, when we unfold this black box. The figure shows how the shape of the EKC flattens when effective environmental policies, defined property rights, innovation, and technological progress are present and how these factors are endogenized in the development process. For example, EKC1 traces the path of environmental degradation when technological progress, policies, regulations, and other related barriers in improving environment are not internalized into the growth process. Over a period of time, EKC1 shifts to EKC2 and EKC3 as the economy introduces environmental policies, technology, infrastructure, and other related factors, which reduces pollutants and environmental degradation.

Environmental degradation may reduce the productivity of workers and can indirectly affect the economic output. A simultaneous equation model is more appropriate than reduced-form models used in the earlier literature. Following Stern (2004), earlier studies ignored the statistical properties of the data series (such as serial dependence or stochastic trends in the series) along with omitted variable bias.

In the later phase, additional explanatory variables were used. Investment shares, infrastructure, electricity tariffs, political rights, civil liberties, and trade have been added to the empirical models. However, most of these studies concluded that income was the most significant variable affecting environmental quality. Another general empirical observation was that the turning point for individual pollutants varies across countries. This difference reflects the positive spillover effects of the technology, research, and development of the first movers on the follower countries. In the most recent literature on the EKC, most of the research was based on time series analysis and/or panel estimations identifying cross-sectional dependence and heterogeneity across countries; we refrain from this discussion in the current chapter as it will be discussed in the next chapter.

POLICY PERSPECTIVES AND CONCLUDING REMARKS

The early literature is based on a cross-country approach because of the absence of long time-series of environmental data. Although environmental degradation is generally with upward trend in developing countries, the trend may not be decreasing in many developed countries. The rate of degradation is slower in the developed world and faster in the developing countries. Therefore, these studies may merely reflect the combining effects of these two opposite trends, instead of showing path of environmental degradation followed by a single country over a time period. The studies based on time series data cast serious doubt on the shape of the EKC; therefore, policy advisers should not consider the shape of this curve as a guideline but draw conclusions on the downward trend that occurs automatically in the long run at high-income levels. In middle-income and developing countries, the structural composition and technology play a greater role than economies of scale for improving environmental quality with rising income; local and institutional reforms (e.g., environmental legislations and market-based incentives) play a greater role in reducing emissions.

The height or degree of the convexity of the EKC depends on policies and markets within an economy. The presence of effective environmental policies, energy subsidies, and defined property rights will reduce environmental degradation and the height of the EKC for any income level; the speed of adjustment, both in economic and social changes, is significant in this respect. Abatement effects at higher income levels, along with structural changes within economies, will play a greater role in defining the trajectory of environmental quality.

Since the earliest studies, literature on the EKC has grown remarkably. Despite the detailed data, relevant key variables, and new econometric techniques used for estimating the models, there is still no universal shape of the EKC across countries. The major conclusions we can draw from early literature are as follows:

- A universal shape of the EKC is nonexistent. The shape varies with different environmental pollutants within a country, across countries, and with different econometric techniques, researchers use to analyze the income-environmental pollutant nexus.
- Growth and pollution trajectories are different across countries. Growth and pollution trajectories are interrelated within a country. Country-specific studies with the application of time series analysis are needed; this is important for developing countries where environmental degradation is a major problem.
- In following the stages of development, developing countries cannot follow the path of developed countries. Policies for sustainable development should be implemented in the early stages of development. In other words, a country should not wait until the turning point to occur before reaping the benefits of research, innovation, and effective environmental policies to reduce environmental degradation.
- Implementation of research, innovation, and various environmental policies are needed to endogenize the negative externality of environmental resources into the growth process.

REFERENCES

Azomahu, T., & Van Phu, N. (2001). *Economic growth and CO$_2$ emissions: A nonparametric approach.* BETA Working Paper N.2001-01.

Barbier, E. B. (1997). Introduction to the environmental Kuznets curve special issue. *Environment and Development Economics, 2*(4), 369−381.

Beckerman, W. (1992). Economic growth and the environment: Whose growth? whose environment? *World Development, 20*(4), 481−496.

Carson, R. T., Jeon, Y., & McCubbin, D. R. (1997). The relationship between air pollution emissions and income: US data. *Environment and Development Economics, 2*(4), 433−450.

Cole, M. A. (2004). Trade, the pollution haven hypothesis and the environmental Kuznets curve: Examining the linkages. *Ecological Economics, 48*(1), 71−81, 2004.

Cole, M. A., Rayner, A. J., & Bates, J. M. (1997). The environmental Kuznets curve: An empirical analysis. *Environment and Development Economics, 2*(4), 401−416.

De Bruyn, S. M., van den Bergh, J. C., & Opschoor, J. B. (1998). Economic growth and emissions: Reconsidering the empirical basis of environmental Kuznets curves. *Ecological Economics, 25*(2), 161−175.

Dijkgraaf, E., & Vollebergh, H. R. (2001). *A note on testing for environmental Kuznets curves with panel data (No. 63.2001)*. Nota di Lavoro, Fondazione Eni Enrico Mattei.

Dinda, S. (2004). Environmental Kuznets curve hypothesis: A survey. *Ecological Economics, 49*(4), 431–455.

Dinda, S., & Coondoo, D. (2006). Income and emission: A panel data-based cointegration analysis. *Ecological Economics, 57*(2), 167–181.

Galeotti, M., Lanza, A., & Pauli, F. (2006). Reassessing the environmental Kuznets curve for CO2 emissions: A robustness exercise. *Ecological Economics, 57*(1), 152–163.

Grossman, G. M., & Krueger, A. B. (1991). *Environmental impacts of a North American free trade agreement (No. w3914)*. National Bureau of Economic Research.

Grossman, G., & Krueger, A. (1995). Economic growth and the environment. *Quarterly Journal of Economics, 110*, 353–377.

Holtz-Eakin, D., & Selden, T. M. (1995). Stoking the fires? CO_2 emissions and economic growth. *Journal of Public Economics, 57*(1), 85–101.

Kuznets, S. (1955). Economic growth and income inequality. *The American Economic Review, 45*(1), 1–28.

Kuznets, S. (1998). Population, income and capital. *International Social Science Journal, 50*(157), 329–334.

Martínez-Zarzoso, I., & Bengochea-Morancho, A. (2004). Pooled mean group estimation of an environmental Kuznets curve for CO_2. *Economics Letters, 82*(1), 121–126.

McConnell, K. E. (1997). Income and the demand for environmental quality. *Environment and Development Economics, 2*(4), 383–399.

Meadows, D. H., Meadows, D. L., Randers, J., & Behrens, W. (1972). *The limits to growth*. New York, USA: Universe Books.

Panayotou, T. (1993). *Empirical tests and policy analysis of environmental degradation at different stages of economic development (No. 992927783402676)*. International Labour Organization.

Pearson, P. J. (1994). *Energy, externalities and environmental quality: Will development cure the ills it creates?* Surrey Energy Economics Centre, Department of Economics, University of Surrey.

Richmond, A. K., & Kaufmann, R. K. (2006). Is there a turning point in the relationship between income and energy use and/or carbon emissions? *Ecological Economics, 56*(2), 176–189.

Roberts, J. T., & Grimes, P. E. (1997). Carbon intensity and economic development 1962–91: A brief exploration of the environmental Kuznets curve. *World Development, 25*(2), 191–198.

Schmalensee, R., Stoker, T. M., & Judson, R. A. (1998). World carbon dioxide emissions: 1950–2050. *The Review of Economics and Statistics, 80*(1), 15–27.

Shafik, N., & Bandyopadhyay, S. (1992). *Economic growth and environmental quality: Time series and cross-country evidence*. Washington, DC: The World Bank. Background Paper for the World Development Report 1992.

Stern, D. I. (2004). The rise and fall of the environmental Kuznets curve. *World Development, 32*(8), 1419–1439.

Stern, D. I., Common, M. S., & Barbier, E. B. (1996). Economic growth and environmental degradation: The environmental Kuznets curve and sustainable development. *World Development, 24*(7), 1151–1160.

Suri, V., & Chapman, D. (1998). Economic growth, trade and the environment: Implications for the environmental Kuznets curve. *Ecological Economics, 25*, 195–208.

Taskin, F., & Zaim, O. (2000). Searching for a Kuznets curve in environmental efficiency using Kernel estimation. *Economics Letters, 68*, 217–223.

Vincent, J. R. (1997). Testing for environmental Kuznets curves within a developing country. *Environment and Development Economics, 2*(4), 417–431.

World Bank. (1992). *World development report 1992*. New York: Oxford University Press.

Recent Studies (Extending Basic Environmental Kuznets Curve Model by Adding More Variables)

ROULA INGLESI-LOTZ, PHD

RECENT STUDIES (EXTENDING BASIC ENVIRONMENTAL KUZNETS CURVE MODEL BY ADDING MORE VARIABLES)

As discussed extensively thus far, the relationship between environmental degradation and income levels or economic growth demonstrated within the EKC hypothesis is a rather complex one. Even though the literature has shown particular interest in examining the EKC hypothesis, consensus on its existence and characteristics has not been reached. Results in various studies might differ due to their methodologic approaches, time periods, and the inclusion (or exclusion thereof) of important variables that might affect the relationship (Alam, Murad, Noman, & Ozturk, 2016; Inglesi-Lotz & Pouris, 2016). Karanfil (2009) also points that not only endogenous but also exogenous factors can influence the relationship between emissions (or energy) and economic growth. Multivariate models are preferable because they can capture case-specific factors for different sectors and geographical regions (Ozturk, 2010).

This chapter presents some of those factors that have been included in various recent studies to capture the EKC in a more robust way. A few studies will be discussed for each of these factors along with the theoretic framework and underpinnings (particularly how each of these factors influence economic growth and emissions).

Studies Including Tourism Variables

Among recent globalization trends, tourism development and activities created strong links between countries. The tourism sector has the capacity of creating foreign exchange earnings, infrastructural development, job creation, and revenues from international visitors. In addition, its great potential to boost economic growth is derived by the inclusion of a suite of business sectors such as hospitality, transport, etc., as well as its ability to attract more investment in human and physical capital (Wang and Wang (2018); Fayissa, Nsiah, & Tadasse, 2008). The relationship of tourism and economic growth has been summarized within two hypotheses in the literature: tourism-led growth and growth-led tourism (Pablo-Romero & Molina, 2013). A number of studies have found that tourism has a direct and positive causal effect on economic growth (Balaguer & Cantavella-Jorda, 2002; Dritsakis, 2004, 2012; Durbarry, 2002; Eugenio-Martin, Martin Morales, & Scarpa, 2004; Gunduz & Hatemi-J., 2005; Narayan, 2004). On the other side, some studies have advocated the notion that economic growth affects positively the development of the tourism sector: when an economy growth steadily, usually infrastructure, health and security improve as well and thus attract international tourism but also encourage national tourism (Narayan, 2004; Oh, 2005; Tang & Jang, 2009).

Regardless of all economic benefits of the tourism sector, its economic activities cause environmental degradation (as most economic activities), such as air and marine pollution, soil erosion, and others. The development of the specific sector needs infrastructure building such as airports and road networks as well as a variety of services (hotels, restaurants, shops, and others) (Ozturk, Al-Mulali, & Saboori, 2016). Qureshi et al. (2017) explain that tourism at initial stages of development influences the environmental quality of a country. However, at the next stages, tourism growth promotes an inverted U-shaped EKC hypothesis trend that assists in promotion of sustainable policies in the tourism sector (De Vita et al., 2015).

De Vita et al. (2015) explain that the inclusion of tourism in a traditional EKC model will influence the size of the scale, composition, and technique effects. "To gauge whether despite such a direct influence on environmental degradation, further investment in tourism development could—by fueling further economic growth—help the economy reach the exponential growth levels at which CO_2 emissions would be expected to decline" (De Vita et al., 2015, p. 16654).

As the call for reduction of greenhouse gas emissions intensify lately, a number of studies have started examining the relationship between environmental indicators and tourism development through the lens of economic growth and the EKC hypothesis (De Vita et al., 2015; Dogan, 2015, Dogan, Seker, & Bulbul, 2015). Katircioglu, Feridun, and Kilinc (2014) applied the ARDL model to examine the relationship among emissions, energy consumption, and tourism for the economy of Cyprus from 1970 to 2009. They conclude that tourism has a positive impact on CO_2 emissions and energy consumption. The same results are confirmed by Solarin (2014) for Malaysia and Katircioglu (2014) for Turkey. Various studies examined the nexus in multicountry cases. For example, Dogan et al. (2015) found a there is a unidirectional causality from tourism to CO_2 emissions for the OECD countries, whereas Zaman, Shahbaz, Loganathan, and Raza (2016) also concluded a causality from tourism to emissions for a panel of East Asia and Pacific, the European Union, and high-income OECD and non-OECD countries. Dogan and Aslan (2017) confirm the findings for the EU countries, whereas Paramati, Shahbaz, and Alam (2017) agree for the case of Eastern and Western EU countries. Although most studies have found a positive impact of tourism to CO_2 emissions, Lee and Brahmasrene (2013) found that although tourism has a positive influence on GDP, it has a negative effect on CO_2 emissions.

Although mostly expecting negative impacts to be found from tourism to the environment, some studies have also advocated that damages in the environment might have a negative effect to attraction of tourists (Giannoni & Maupertuis, 2007). Studies such as Huybers and Bennett (2000) showed that air pollution influences the selection of destination choices of tourists and Zhang, Zhong, and Xu (2015) conclude that haze pollution makes Beijing an unattractive destination for international tourism.

The magnitude of impact of tourism activities to a country's environmental profile highly depends on the choice of energy type, whether renewable or nonrenewable. In addition, it may differ based on the variety of policies and government interventions and the use of clean technologies (Dogan & Aslan, 2017). On the use

of appropriate policies and interventions, Cerina (2007) has demonstrated that tourism development can be conductive to better environmental quality and economic conditions only through under certain assumptions and can be maintained only "when pollution abatement expenditures are no longer available" (Wang & Wang, 2018). Proposed solutions to the environmental degradation caused by tourism development are disinvestment to the sector (Giannoni & Maupertuis, 2007) and implementation of a tax on revenue (Alfsen, Birkelund, and Aaserud (1995); Smeral, 1996). Wunder (2000) also proposed the idea of ecotourism promotion as a solution through nature conservation, to which Das and Chatterjee (2015) agree but stressing that "proper monitoring, evaluation, and management."

Studies Including Financial Development

Long before examining the effects of financial development to the emission levels globally, Schumpeter (1911) set the foundations for the relationship between financial development and economic growth: a well-developed financial system that mobilizes savings and provides access to capital for investment, enhancing domestic production by allocating funds accordingly. Shahbaz (2009) also discusses the importance of the sector in reducing transaction, information, and monitoring costs. Shahbaz (2013) adds that the "financial sector also encourages investment activities by issuing loans at a cheaper cost and allocates resources to productive ventures, mobilizes savings, enabling trading, offering hedging, diversifying the risks, monitoring the working of the firms, and directs the firm to use environment friendly technology to enhance the level of domestic output."

Halkos and Sepetis (2007), He and Wang (2012), and Halkos and Polemis (2016) acknowledge the lack of consensus in the literature that deals with the relationship between environmental pollution and financial development. According to Frankel and Romer (1999), the process of financial development promotes adoption of cleaner technologies and thus reduce the environmental effects of higher growth; however, studies by Zhang (2011) and Shahbaz and Lean (2012) conclude that through stimulation of economic activity, efficiency in the stock market, and attraction of Foreign Direct Investment (FDI) lead to increases in emissions. Bello and Adimbola (2010) concluded that financial development, measured by stock market capitalization, contributed to the emissions of Nigeria. This intensification of emissions was attributed to the lack of loan monitoring to investment projects.

Developing an accessible and sophisticated financial sector can promote the adoption of environmental-friendly technologies (Birdsall & Wheeler, 1993; Frankel & Rose, 2002) and thus, contribute in the reduction of greenhouse gas emissions. In addition, a developed financial sector has the potential to lower emissions by higher efficiency in the energy sector as well as promotion of technologic innovations (Tamazian & Rao, 2010; Tamazian, Pineiro, & Vadlamannati, 2009). Low-borrowing costs also assist national and local governments to embark on environmental projects (Shahbaz, 2013). For example, Tamazian et al. (2009) studied the relationship between economic and financial development and CO_2 emissions for the United States, Japan, and the BRIC countries confirming that both economic and financial growth has an impact in reducing CO_2 emissions.

Finally, studies have also at times shown a lack of relationship between financial development and emission levels. Ozturk and Acaravci (2013) found that financial development had no impact to CO_2 emissions levels in Turkey; whereas Lee, Chen, and Cho (2015) reject the EKC hypothesis for a panel of 25 OECD countries for the period of 1971 to 2007 with the inclusion of financial development.

Studies Using Urbanization and Population as Determinants of Pollution

With the continuous progression of economies, globalization, and concentration of employment opportunities in urban areas, main cities have become the centers of economic activity, politics, education, and culture (Lin & Zhu, 2018). These trends and population movements described as *urbanization* pose serious concerns concerning energy demand and management, demand for infrastructure, as well as air pollution in the form of emissions (Jiang & Lin, 2012). Especially, the speed of urbanization recently has affected the already restricted availability of resources with negative consequences to environmental conditions of countries (Yang, Huang, & Thompson, 2014; Li et al., 2018).

However, at the same time, the effect of urbanization on GDP per capita is generally positive as the individuals moving to the urban centers have higher chances of being used and increase their income levels and hence economic development (Yang, Liu, & Zhang, 2017; Jing, 2015). Behera and Dash (2017, p. 96) stress further "the trend of urbanization has been historically suggested as the precondition for development in the developing world. Starting from the industrial revolution in the west it has been significantly accepted that rationale behind the industrial-urban inter-linkages can be the pathfinder to the growth and subsequent economic wellbeing of the society."

Taking into consideration these dynamics (refer to Fig. 3.1), the literature on the importance of urbanization in the emissions—economic growth nexus, represented by the EKC hypothesis, is not small. The study by Glaeser and Kahn (2010) is considered among the

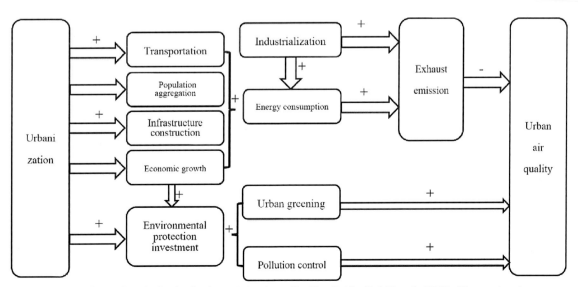

FIG. 3.1 Dynamics of urbanization impacts to air quality. (Credit: Lin, B. & Zhu, J., (2018). Changes in urban air quality during urbanization in China. *Journal of Cleaner Production*, 188, 312–321.)

pioneers in the debate of urbanization and environmental conditions. As the characteristics of urbanization evolve over time, the need of examine the topic is continuous (Grimm et al., 2008; Shao, Tang, Zhang, & Li, 2006). Studies have advocated in the past that increase in the share of urban population to total has worsened the environmental quality of cities (Song, Deng, & Liu, 2015). The concentration of population at cities has created higher demand for infrastructure construction, transportation, and production of goods (Sun, Ouyang, & Cai, 2014; Du, Lin, & Sun, 2015). The rising energy consumption, depending on the type of energy of course, worsens the environmental conditions (Satterthwaite, 2009; Shafiei & Salim, 2014; Rafiq, Salim, & Nielsen, 2016; Cole & Neumayer, 2004). Parikh and Shukla (1995) conclude that urbanization increases greenhouse gas emissions for 83 countries, whereas York, Rosa, and Dieta (2003) agree with the results for country group of 137 countries.

On the other side, the improved economic conditions created will assist people have higher environmental standards and promote greener policies, making thus urbanization a factor accelerating the EKC phenomenon (Lin & Zhu, 2018). Martinez-Zarzoso and Maruotti (2011) examined a group of developing countries and found an inverted U-shaped relationship between emissions and urbanization. Wang, Ma, and Zhao (2016) and Shahbaz, Loganathan, and Muzaffar (2016) also confirmed the hypothesis of an inverted U-shaped relationship. Al-Mulali et al. (2013) concluded a bidirectional relationship between urbanization and CO_2 emissions in the MENA countries.

There are also studies that support the perspective that the urbanization process, if done properly, will promote better land and resources allocation and centralized policies that will assist in mitigation of greenhouse gas emissions (Zeng & Zhao, 2009). For developing countries, Fan, Lui, Wu, and Wie (2006) and Ala-Mantila et al. (2014) found that urbanization could decrease carbon dioxide emissions, whereas Liddle and Lung (2010) showed that urbanization does not have an impact on total greenhouse gas emissions for developed countries. Effiong (2017) and Li and Zhou (2017) both have concluded that urbanization is a means to reduce environmental degradation in African countries and China, respectively.

Studies have examined specifically the impact of urbanization to environmental quality through transport infrastructure demand. Sun, Luo, and Li (2018) showed that it overall improves pollution: in the short term, traffic and concentration of population will be the reason for a rise in pollutants, but in the long run, investment in transport infrastructure to accommodate rising urban populations might improve network and public transport so much so that pollution levels decrease. Poumanyvong, Kaneko, and Dhakal (2012) also find that urbanization trends tend to improve a country's public transport and network, effectively improving the environmental quality of the country.

Overall, the recent urbanization waves have urged research on their potential impacts to economic development and environmental quality. The results in the literature are mixed resulting from examining different perspectives time-wise—Poumanyvong and Kaneko (2010) explain that the magnitude of the urbanization's impact depends on the specific developmental stage of the economy. Sadorsky (2014) suggests from the findings that there are simultaneously two opposite effects that cancel each other out.

Where all studies agree conclusively is the necessity of coordinated planning and policies to absorb the movement of population, proper city development, as well as future construction of infrastructure in rural areas to catch up with urban development (Lin & Zhu, 2018). Behera and Dash (2017) also propose the concept of sustainable urbanization: the development of smart and green policies to internalize the negative externalities of rising urbanization.

Studies Using Energy Consumption Variables

Energy consumption is the major factor linking the two variables considered when testing the EKC hypothesis in both single- and multiple-country cases. Greenhouse gas emissions are considered in their big majority (and mainly the CO_2) a result of human activities, such as burning fossil fuels, extreme air pollution from the industrial sector, and high rates of deforestation (Danish, Zhang, Wang, & Wang, 2017).

At the same time, a number of studies in the literature have advocated that economic growth and development cannot be achieved without energy consumption (Ahmed, Shahbaz, Qasim, & Long, 2015; Chadran & Tang, 2013; Chow & Li, 2014; Fahrani & Ozturk, 2015; Hamit-Haggar, 2012; Kohler, 2013; Lau, Choong, & Eng, 2014; Saboori & Sulaiman, 2013; Shahbaz, Hooi, & Shahbaz, 2012; Yavuz, 2014). As Ozokcu and Ozdemir (2017) stress, "energy is a fundamental input for not only production but also consumption; that is, it is the basis of economic growth." Jobert, Karanfil, and Tykhonenko (2014) also link the energy with severe environmental consequences although admits its importance as a production factor. Ghali and El-Sakka (2004) in Inglesi-Lotz (2018) express two main but opposite perspectives: (1) no production factor can function without energy and hence its importance and (2) relatively to other factors, energy costs are still low proportionally to total economic production.

However, in the argumentation of including an energy variable to capture some of the EKC dynamics, some studies would argue that the type of energy is crucial. In a country's supply mix that is dominated by fossil fuels, the effects to the emissions would be more severely negatively that in a cleaner supply mix. Hence, some studies, since the 2000s, started including the consumption of specific energy source in their analysis of the EKC hypothesis; particularly, fossil fuel consumption is found to be a primary positive contributor of greenhouse gas emissions (Apergis & Payne, 2009; Halicioglu, 2009; Kanjilal & Ghosh, 2013; Rafindadi, 2016; Riti, Song, Shu, & Kamah, 2017; Shahbaz, Khraief, Uddin, & Ozturk, 2014).

Focusing on the renewable energy consumption and its effect in the inverted U-shaped relationship between emissions and GDP, Boluk and Mert (2015) examined the Turkish case and confirmed the EKC hypothesis, where they found that renewable energy slowed down the emissions of the country. Cerdeira Bento and Moutinho (2016) aimed at examining the potential impact of both renewable and nonrenewable energies and even with the inclusion of both; the EKC was confirmed in Italy. The results of Baek (2016) agree too for the case of the US economy. Sulaiman, Azman, and Saboori (2013) also confirm the EKC hypothesis for Malaysia and confirm that the use of renewable energy reduces the country's emissions. Within a panel framework, Al-Mulali, Ozturk, and Adebola (2016) investigated the EKC hypothesis with the inclusion of renewable energy within seven world regions: it was confirmed for Europe, South Asia, and America but not for the Gulf states, North Africa, and sub-Saharan Africa. Zoundi (2017) also makes a point that "renewable energy, with a negative effect on CO_2 emissions, coupled with an increasing long-run effect, remains an efficient substitute for the conventional fossil-fueled energy" for 25 African countries.

Continuing with the motivation for clean energy options maximized in the supply mix, several studies examined the role of nuclear energy in the EKC context. Richmond and Kaufmann (2006) examined the inclusion of nuclear energy in the analysis using panel data for OECD and non-OECD countries. Apergis, Payne, Menyah, and Wolde-Rufael (2010) concluded a causal relationship from nuclear energy consumption to

emissions for a group of 19 countries and agreed also in that by Iwata, Okada, and Samreth (2010) for France, Menyah and Wolde-Rufael (2010) for the United States, Iwata, Okada, and Samreth (2012) for OECD countries, Baek (2015) for major nuclear generation countries, and Ozturk (2017) for Latin American countries. For the United States and nine developed countries, although, Jaforullah and King (2015) and Saidi and Mbarek (2016) concluded no causality from nuclear energy consumption to greenhouse gas emissions.

Conclusions and Policy Implications

As discussed in the beginning of the chapter, the literature has not managed to reach a consensus with regard to the existence of the inverted U-shaped type of relationship between emissions and income per capita, as described by the EKC hypothesis. There are multiple reasons behind this lack of agreement, such as the differences in methodology, time periods of the investigation, and the country or countries examined.

Individual countries as well as group of countries have particular socioeconomic and environmental characteristics that potentially affect the relationship between emissions and economic growth, within the EKC context. This chapter examines some of these factors and their impact toward examining the EKC hypothesis: tourism activities, financial development, urbanization and population, and energy consumption. Interestingly, all of these factors affect both emissions and economic growth through different channels.

The inclusion of additional variables is important not only for the modeling exercise but also from a policy perspective. Countries and policymakers have made commitments to the international community to decrease emissions in the following decades, but at the same time, this should not be done at the expense of economic growth and development or the welfare of the global population.

Hence, the ultimate goal is to decouple economic growth from greenhouse gas emissions and within the EKC framework, be on the right-hand side of the curve, where income per capita and emissions have a negative relationship. Policymakers pursue various avenues to activate this direction of the relationship. Whether environmental conditions improve as well as when and how is highly dependent on policies, social institutions as well as the optimal operation of markets. It is crucial thus for these policies to be informed appropriately with regard to all the determinants that can affect their success and the economic sectors and activities that are influential to not only decoupling the relationship between emissions and economic growth but also potentially to promote reduction of emissions when economies develop and grow.

Improvement of the environment with income growth is not automatic but depends on policies and institutions. GDP growth creates the conditions for environmental improvement by raising the demand for improved environmental quality and makes the resources available for supplying it. Whether environmental quality improvements materialize or not, and when and how, depend critically on government policies, social institutions, and the completeness and functioning of markets.

Potential limitation in the technical analysis of EKC when other variables are included is the issue of endogeneity. All the determinants in this chapter are highly correlated with GDP measuring economic growth—so their inclusion as determinants of emissions along with GDP might skew the results. Another possible problem may arise within the specifics of the methodologic approach. Some of these determinants might exhibit an asymmetric impact to emissions: stronger in magnitude during high economic development, for example, than during low economic development. Therefore, econometric and other methods that take into account the regime—switching behaviors when investigating the specific nature of the EKC hypothesis are represented in a two-sided graph: where on the one side, the economy is growing, mainly using dirty industries, and the other side, the economy is well established and mainly growing using more environmentally friendly industries.

REFERENCES

Ahmed, K., Shahbaz, M., Qasim, A., & Long, W. (2015). The linkages between deforestation, energy, and growth for environmental degradation in Pakistan. *Ecological Indicators, 49*, 95−103.

Al-Mulali, U., Fereidouni, H. G., Lee, J. Y., & Sab, C. B. (2013). Exploring the relationship between urbanization, energy consumption, and CO_2 emissions in MENA countries. *Renewable and Sustainable Energy Reviews, 23*, 107−112.

Al-Mulali, U., Ozturk, I., & Adebola, S. (2016). Investigating the environmental Kuznets curve hypothesis in seven regions: The role of renewable energy. *Ecological Indicators, 67*, 267−282.

Ala-Mantila, S., Heinonen, J., & Junnila, S. (2014). Relationship between urbanization, direct and indirect greenhouse gas emissions, and expenditures: A multivariate analysis. *Ecological Economics, 66*, 129−139.

Alam, M., Murad, W., Noman, A., & Ozturk, I. (2016). Relationships among carbon emissions, economic growth, energy consumption and population growth: Testing environmental Kuznets curve hypothesis for Brazil, China, India and Indonesia. *Ecological Indicators, 70*, 466−479.

Alfsen, K. H., Birkelund, H., & Aaserud, M. (1995). Impacts of an EC carbon/energy tax and deregulating thermal power supply on CO_2, SO_2, and NOx emissions. *Environmental and Resource Economics, 5*, 165−189.

Apergis, N., & Payne, J. E. (2009). CO_2 emissions, energy usage, and output in Central America. *Energy Policy, 37*, 3282−3286.

Apergis, N., Payne, J. E., Menyah, K., & Wolde-Rufael, Y. (2010). On the causal dynamics between emissions, nuclear energy, renewable energy, and economic growth. *Ecological Economics, 69*, 2255−2260.

Baek, J. (2015). A panel cointegration analysis of CO_2 emissions, nuclear energy, and income in major nuclear generating countries. *Applied Energy, 145*, 133−138.

Baek, J. (2016). Do nuclear and renewable energy improve the environment? Empirical evidence from the United States. *Ecological Indicators, 66*, 352−356.

Balaguer, J., & Cantavella-Jorda, M. (2002). Tourism as a a longrun economic growth factor: The Spanish case. *Applied Economics, 34*, 877−884.

Behera, S. R., & Dash, D. P. (2017). The effect of urbanization, energy consumption, and foreign direct investment on the carbon dioxide emission in the SSEA (South and Southeast Asian) region. *Renewable and Sustainable Energy Reviews, 70*, 96−106.

Bello, A. K., & Adimbola, O. M. (2010). Does the level of economic growth influence environmental quality in Nigeria: A test of environmental Kuzents curve (EKC) hypothesis. *Pakistan Journal of Social Sciences, 7*, 325−329.

Birdsall, N., & Wheeler, D. (1993). Trade policy and industrial pollution in Latin America: Where are the pollution havens? *Journal of Environment and Development, 2*, 137−149.

Boluk, G., & Mert, M. (2015). The renewable energy, growth and environmental Kuznets curve in Turkey: An ARDL approach. *Renewable and Sustainable Energy Reviews, 52*, 587−595.

Cerdeira Bento, J. P., & Moutinho, V. (2016). CO_2 emissions, non-renewable and renewable electricity production, economic growth, and international trade in Italy. *Renewable and Sustainable Energy Reviews, 55*, 142−155.

Cerina, F. (2007). Tourism specialization and environmental sustainability in a dynamic economy. *Tourism Economics, 13*, 553−582.

Chadran, V. G., & Tang, C. F. (2013). The impacts of transport energy consumption, foreign direct investment and income on CO_2 emissions in ASEAN- 5 economies. *Renewable and Sustainable Energy Reviews, 24*, 445−453.

Chow, G. C., & Li, J. (2014). Environmental Kuznets curve: Conclusive econometric evidence for CO_2. *Pacific Economic Review, 19*, 1−7.

Cole, M. A., & Neumayer, E. (2004). Examining the impact of demographic factors on air pollution. *Population and Environment, 26*, 5−21.

Danish, Zhang, B., Wang, B., & Wang, Z. (2017). Role of renewable energy and non-renewable energy consumption on EKC: Evidence from Pakistan. *Journal of Cleaner Production, 156*, 855−864.

Das, M., & Chatterjee, B. (2015). Ecotourism: A panacea or a predicament. *Tourism Mamangement perspectives, 14,* 3−16.

De Vita, G., et al. (2015). Revisiting the environmental Kuznets Curve hypothesis in a tourism development context. *Environmental Science and Pollution Research, 22,* 16652−16663.

Dogan, E. (2015). The relationship between economic growth and electricity consumption from renewable and non-renewable sources: A study of Turkey. *Renewable and Sustainable Energy Reviews, 52,* 534−546.

Dogan, E., & Aslan, A. (2017). Exploring the relationship among CO_2 emissions, real GDP, energy consumption and tourism in the EU and candidate countries: Evidence from panel models robust to heterogeneity and cross-sectional dependence. *Renewable and Sustainable Energy Reviews, 77,* 239−245.

Dogan, E., Seker, F., & Bulbul, S. (2015). Investigating the impacts of energy consumption, real GDP, tourism and trade on CO_2 emissions by accounting for cross-sectional dependence: A panel study of OECD countries. *Current Issues in Tourism,* 1−19.

Dritsakis, N. (2004). Tourism as a long-run economic growth factor: An empirical investigation for Greece using causality analysis. *Tourism Economics, 10,* 305−316.

Dritsakis, N. (2012). Tourism development and economic growth in seven mediterranean countries: A panel data approach. *Tourism Economics, 18,* 801−816.

Du, G., Lin, W., & Sun, C. (2015). Residential electricity consumption after the reform of tiered pricing for household electricity in China. *Applied Energy, 157,* 276−283.

Durbarry, R. (2002). The economic contribution of tourism in Mauritius. *Annals of Tourism Research, 29,* 862−865.

Effiong, E. L. (2017). On the urbanization-pollution nexus in Africa: A semiparametric analysis. *Quality and Quantity, 52,* 1−12.

Eugenio-Martin, J. L., Martin Morales, N., & Scarpa, R. (2004). *Tourism and economic growth in Latin American countries: A panel data approach.* Social Science Research Network. s.l.: Working Paper 26.

Fahrani, S., & Ozturk, I. (2015). Causal relationship between CO_2 emissions, real GDP, energy consumption, financial development, trade openness, and urbanization in Tunisia. *Environmental Science and Pollution Research, 22,* 15663−15676.

Fan, Y., Lui, L. C., Wu, G., & Wie, Y. M. (2006). Analyzing impact factors of CO_2 emissions using the STIRPAT model. *Environmental Impact Assessment, 26,* 377−395.

Fayissa, B., Nsiah, C., & Tadasse, B. (2008). Impact of tourism on economic growth and development in Africa. *Tourism Economics, 14,* 807−818.

Frankel, J., & Romer, D. (1999). Does trade cause growth? *The American Economic Review, 89,* 379−399.

Frankel, J., & Rose, A. (2002). An estimate of the effect of common currencies on trade and income. *Quarterly Journal of Economics, 117,* 437−466.

Ghali, K. H., & El-Sakka, M. I. (2004). Energy use and output growth in Canada: A multivariate cointegration analysis. *Energy Economics, 26,* 225−238.

Giannoni, S., & Maupertuis, M. A. (2007). Environmental quality and optimal investment in tourism infrastructure: A small island perspective. *Tourism Economics, 13,* 499−513.

Glaeser, E. L., & Kahn, M. E. (2010). The greenness of cities: Carbon dioxide emissions and urban development. *Journal of Urban Economics, 67,* 404−418.

Grimm, N. B., et al. (2008). Global change and the ecology of cities. *Science, 319,* 756−760.

Gunduz, L., & Hatemi-J, A. (2005). Is tourism-led growth hypothesis valid for Turkey. *Applied Economics Letters, 12,* 499−504.

Halicioglu, F. (2009). An econometric study of CO_2 emissions, energy consumption, income and foreign trade in Turkey. *Energy Policy, 37,* 1156−1164.

Halkos, G., & Polemis, M. (2016). *Examining the impact of financial development on the environmental Kuznets curve hypothesis.* s.l.: Munich Personal RePEc Archive (MPRA) No 75368.

Halkos, G., & Sepetis, A. (2007). Can capital market respond to environmental policy of firms? Evidence from Greece. *Ecological Economics, 63,* 578−587.

Hamit- Haggar, M. (2012). Greenhouse gas emissions, energy consumption and economic growth: A panel cointegration analysis from Canadian industrial sector perspective. *Energy Economics, 34,* 358−364.

He, J., & Wang, H. (2012). Economic structure, development policy and environmental quality: An empirical analysis of environmental Kuznets curves with Chinese municipal data. *Ecological Economics, 76,* 49−59.

Huybers, T., & Bennett, J. (2000). Impact of the environment on holiday destination choices of prospective UK tourists: Implications for tropical north Queensland. *Tourism Economics, 6,* 21−46.

Inglesi-Lotz, R. (2018). The role of potential factors/actors and regime switching modeling. In A. Menegaki (Ed.), *The Economics and Econometrics of the energy-growth nexus* (pp. 113−139). Elsevier. s.l.

Inglesi-Lotz, R., & Pouris, A. (2016). On the causality and determinants of energy and electricity demand in South Africa: A review. *Energy Sources, Part B: Economics, Planning and Policy, 11,* 626−636.

Iwata, H., Okada, K., & Samreth, S. (2010). Empirical study on the environmental Kuznets curve for CO_2 in France: The role of nuclear energy. *Energy Policy, 38,* 4057−4063.

Iwata, H., Okada, K., & Samreth, S. (2012). Empirical study on the determinants of CO_2 emissions: Evidence from OECD countries. *Applied Economics, 44,* 3513−3519.

Jaforullah, M., & King, A. (2015). Does the use of renewable energy sources mitigate CO_2 emissions? A reassessment of the US evidence. *Energy Economics, 49,* 711−717.

Jiang, Z., & Lin, B. (2012). China's energy demand and its characteristics in the industrialization and urbanization process. *Energy Policy, 49,* 608−615.

Jing, H. (2015). Empirical study on the urbanization promoting the Chinese economic growth. s.l. In *International conference on computer science and electronic technology.*

Jobert, T., Karanfil, F., & Tykhonenko, A. (2014). Estimating country-specific environmental Kuznets curves from panel data: A Bayesian shrinkage approach. *Applied Economics, 46,* 1449−1464.

Kanjilal, K., & Ghosh, S. (2013). Environmental Kuznets curve for India: Evidence from tests for cointegration with unknown structural breaks. *Energy Policy, 56,* 509−515.

Karanfil, F. (2009). How many times again will we examine the energy-income using a limited range of traditional econometric tools. *Energy Policy, 37,* 1191−1194.

Katircioglu, S. T. (2014). International tourism, energy consumption, and environmental pollution: The case of Turkey. *Renewable and Sustainable Energy Reviews, 36,* 180−187.

Katircioglu, S. R., Feridun, M., & Kilinc, C. (2014). Estimating tourism induced energy consumption and CO_2 emissions: The case of Cyprus. *Renewable and Sustainable Energy Reviews, 29,* 634−640.

Kohler, M. (2013). CO_2 emissions, energy consumption, income and foreign trade: A South African perspective. *Energy Policy, 63,* 1042−1050.

Lau, L. S., Choong, C. K., & Eng, Y. K. (2014). Investigation of the environmental Kuznets curve for carbon emissions in Malaysia: Do foreign direct investment and trade matter? *Energy Policy, 68,* 490−497.

Lee, J. W., & Brahmasrene, T. (2013). Investigating the influence of tourism on economic growth and carbon emissions: Evidence from panel analysis of the European Union. *Tourism Management, 38,* 69−76.

Lee, J. M., Chen, K. H., & Cho, C. H. (2015). The relationship between CO_2 emissions and financial development: Evidence from OECD countries. *The Singapore Ecoonomic Review, 60,* 155.

Liddle, B., & Lung, S. (2010). Age-structure, urbanization, and climate change in developing countries: Revisiting STIRPAT for disaggregated population and consumption related environmental impacts. *Population and Environment, 31,* 317−343.

Li, J., et al. (2018). The effect of urbanization on carbon dioxide emissions efficiency in the Yangtze River Delta, China. *Journal of Cleaner Production, 188,* 38−48.

Lin, B., & Zhu, J. (2018). Changes in urban air quality during urbanization in China. *Journal of Cleaner Production, 188,* 312−321.

Li, Y., & Zhou, J. B. (2017). A study on the impact paths of industrialization and urbanization on urban air quality in China. *Statistical Research, 34.*

Martinez-Zarzoso, I., & Maruotti, A. (2011). The impact of urbanization on CO_2 emissions: Evidence from developing countries. *Ecological Economics, 70,* 1344−1353.

Menyah, K., & Wolde-Rufael, Y. (2010). CO_2 emissions, nuclear energy, renewable energy, and economic growth in the US. *Enegy Policy, 38,* 2911−2915.

Narayan, P. K. (2004). Economic impact of tourism on Fiji's economy: Empirical evidence from the computable general equilibrium model. *Tourism Economics, 10,* 419−433.

Oh, C.-K. (2005). The contribution of tourism development to economic growth in the Korean economy. *Tourism Management, 26,* 39−44.

Ozokcu, S., & Ozdemir, O. (2017). Economic growth, energy, and environmental Kuznets curve. *Renewable and Sustainable Energy Reviews, 72,* 639−647.

Ozturk, I. (2010). A literature survey on energy-growth nexus. *Energy Policy, 38,* 340−349.

Ozturk, I. (2017). Measuring the impact of alternative and nuclear energy consumption, carbon dioxide emissions and oil rents on specific growth factors in the panel of Latin American countries. *Progress in Nuclear Energy, 100,* 71−81.

Ozturk, I., & Acaravci, A. (2013). The long-run and causal analysis of energy, growth, openness and financial development on carbon emissions in Turkey. *Energy Economics, 36,* 262−267.

Ozturk, I., Al-Mulali, U., & Saboori, B. (2016). Investigating the environmental Kuznets curve hypothesis: The role of tourism and ecological footprint. *Environmental Science and Pollution Research, 23,* 1916−1928.

Pablo-Romero, M. D., & Molina, J. A. (2013). Tourism and economic growth: A review of empirical literature. *Tourism Management Perspectives, 8,* 28−41.

Paramati, S. R., Shahbaz, M., & Alam, M. S. (2017). Does tourism degrade environmental quality? A comparative study of Eastern and Western European Union. *Transportation Research Part D, 50,* 1−13.

Parikh, J., & Shukla, V. (1995). Urbanization, energy use and greenhouse effects in economic development: Results from a cross-national study of developing countries. *Global Environmental Change, 5,* 87−103.

Poumanyvong, P., & Kaneko, S. (2010). Does urbanization lead to less energy use and lower CO_2 emissions? A cross-country analysis. *Ecological Economics, 70,* 434−444.

Poumanyvong, P., Kaneko, S., & Dhakal, S. (2012). Impacts of urbanization on national transport and road energy use: Evidence from low, middle and high income countries. *Energy Policy, 46,* 268−277.

Qureshi, M. I., et al. (2017). Dynamic linkages between sustainable tourism, energy, health and wealth: Evidence from top 80 international tourist destination cities in 37 countries. *Journal of Cleaner Production, 158,* 143−155.

Rafindadi, A. A. (2016). Revisiting the concept of environmental Kuznets curve in period of energy disaster and deteriorating income: Empirical evidence from Japan. *Energy Policy, 94,* 274−284.

Rafiq, S., Salim, R., & Nielsen, I. (2016). Urbanization, openness, emissions, and energy intensity: A study of increasingly urbanized emerging economies. *Energy Economics, 56,* 20−28.

Richmond, A. K., & Kaufmann, R. K. (2006). Is there a turning point in the relationship between income and energy use and/or carbon emissions. *Ecological Economics, 56,* 176−189.

Riti, J. S., Song, D., Shu, Y., & Kamah, M. (2017). Decoupling CO_2 emissions and economic growth in China: Is there consistency in estimation results in analyzing environmental Kuznets curve? *Journal of Cleaner Production, 166,* 1448−1461.

Saboori, B., & Sulaiman, J. (2013). CO_2 emissions, energy consumption and economic growth in Association of Southeast Asian Nations (ASEAN) countries: A cointegration approach. *Energy, 55*, 813−822.

Sadorsky, P. (2014). The effect of urbanization on CO_2 emissions in emerging economies. *Energy Economics, 41*, 147−153.

Saidi, K., & Mbarek, M. B. (2016). Nuclear energy, renewable energy, CO_2 emissions, and economic growth for nine developed countries: Evidence from panel granger causality tests. *Progress in Nuclear Energy, 88*, 364−374.

Satterthwaite, D. (2009). The implications of population growth and urbanization for climate change. *Environment and Urbanization, 21*, 545−567.

Schumpeter, J. (1911). *The theory of economic development.* Cambridge, MA: Harvard Univeristy Press.

Shafiei, S., & Salim, R. A. (2014). Non-renewable and renewable energy consumption and CO_2 emissions in OECD countries: A comparative analysis. *Energy Policy, 66*, 547−556.

Shahbaz, M. (2009). A reassessment of finance-growth nexus for Pakistan: Under the investigation of FMOLS and DOLS techniques. *The ICFAI Journal of Applied Economics*, 65−80.

Shahbaz, M. (2013). Does financial instability increase environmental degradation? Fresh evidence from Pakistan. *Economic Modelling, 33*, 537−544.

Shahbaz, M., Hooi, H., & Shahbaz, M. (2012). Environmental Kuznets curve hypothesis in Pakistan: Cointegration and granger causality. *Renewable and Sustainable Energy Reviews, 16*, 2947−2953.

Shahbaz, M., Khraief, N., Uddin, G. S., & Ozturk, I. (2014). Environmental Kuznets curve in an open economy: A bounds testing and causality analysis for Tunisia. *Renewable and Sustainable Energy Reviews, 34*, 325−336.

Shahbaz, M., & Lean, H. H. (2012). Does financial development increase energy consumption? The role of industrialization and urbanization in Tunisia. *Energy Policy, 40*, 473−479.

Shahbaz, M., Loganathan, N., & Muzaffar, A. T. (2016). How urbanization affects CO_2 emissions in Malaysia? The application of STIRPAT model. *Renewable and Sustainable Energy Reviews, 57*, 83−93.

Shao, M., Tang, X., Zhang, Y., & Li, W. (2006). City clusters in China: Air and surface water pollution. *Frontiers in Ecology and the Environment, 4*, 353−361.

Smeral, E. (1996). Economic policy measures for reducing the environmental impact of tourism. *Tourism Economics, 2*, 173−184.

Solarin, S. A. (2014). Tourist arrivals and macroeconomic determinants of CO_2 emissions in Malaysia. *Anatolia, 25*, 228−241.

Song, W., Deng, X., & Liu, B. (2015). Impact of grain for green and grain for blue policies on valued ecosystem services in Shandong province, China. *Advances in Meteorology, 2015*, 1−10.

Sulaiman, J., Azman, A., & Saboori, B. (2013). The potential of renewable energy: Using the environmental Kuznets curve model. *American Journal of Environmental Sciences, 9*, 103−112.

Sun, C., Luo, Y., & Li, J. (2018). Urban traffic infrastructure investment and air pollution: Evidence from the 83 cities in China. *Journal of Cleaner Production, 172*, 488−496.

Sun, C., Ouyang, X., & Cai, H. (2014). Household pathway selection of energy consumption during urbanization process in China. *Energy Conservation Management, 84*, 295−304.

Tamazian, A., Pineiro, J., & Vadlamannati, K. C. (2009). Does higher economic and financial development lead to environmental degradation: Evidence from BRIC countries? *Energy Policy, 37*, 246−253.

Tamazian, A., & Rao, B. B. (2010). Do economic, financial and institutional developments matter for environmental degradation? Evidence from transitional economies. *Energy Economics, 32*, 137−145.

Tang, C., & Jang, S.-C. S. (2009). THe tourism-economy causality in the United States: A sub-industry level examination. *Tourism Management, 30*, 553−558.

Wang, S. J., Ma, H., & Zhao, Y. B. (2016). Exploring the relationship between urbanization and the ecoenvironment − a case of Beijing-Tianjin-Hebel regions. *Ecological Indicators, 45*, 171−183.

Wang, M.-C., & Wang, C.-S. (2018). Tourism, the environment, and energy policies. *Tourism Economics*, 1−18.

Wunder, S. (2000). Ecotourism and economic incentives − an empirical approach. *Ecological Economics, 32*, 465−479.

Yang, H., Huang, X., & Thompson, J. R. (2014). Soil pollution: Urban brownfields. *Science, 344*, 691−692.

Yang, Y., Liu, J., & Zhang, Y. (2017). An analysis of the implications of China's urbanization policy for economic growth and energy consumption. *Journal of Cleaner Production, 161*, 447−461.

Yavuz, N. C. (2014). CO_2 emissions, energy consumption, and economic growth for Turkey: Evidence from a cointegration tset with a structural break. *Energy Sources, Part B: Economics, Planning and Policy, 9*, 229−235.

York, R., Rosa, E. A., & Dieta, T. (2003). STIRPAT, IPAT, ImPACT: Analytic tools for unpacking the driving forces of environmental impacts. *Ecological Economics, 46*, 351−365.

Zaman, K., Shahbaz, M., Loganathan, N., & Raza, S. A. (2016). Tourism development, energy consumption and environmental Kuznets curve: Trivariate analysis in the panel of developed and developing countries. *Tourism Management, 54*, 275−283.

Zeng, D. Z., & Zhao, L. (2009). Pollution havens and industrial agglomeration. *Journal of Environmental Economics and Management, 58*, 141−153.

Zhang, Y. J. (2011). The impact of financial development on carbon emissions: An empirical analysis in China. *Energy Policy, 39*, 2197−2203.

Zhang, A., Zhong, L., & Xu, Y. (2015). Tourists' perception of haze pollution and the potential impacts on travel: Reshaping the features of tourism seasonality in Beijing, China. *Sustainability, 7*, 2397−2414.

Zoundi, Z. (2017). CO_2 emissions, renewable energy and the Environmental Kuznets Curve, a panel cointegration approach. *Renewable and Sustainable Energy Reviews, 72*, 1067−1075.

CHAPTER 4

Single-Country Versus Multiple-Country Studies

ALPER ASLAN, PHD • EYUP DOGAN • BUKET ALTINOZ, PHD CANDIDATE

The number of studies on the validity of the environmental Kuznets curve (EKC) hypothesis seems to have increased significantly in recent years. The results of the studies vary according to the differences in the models used, the types of pollution used, the periods covered, and the countries involved. The studies using time-series methods (single-country) in the literature related to the EKC are reported in Section A of Tables 4.1 and 4.2, whereas the studies using panel methods (panel-country) in the literature are reported in Section B of Tables 4.1 and 4.2. It is worth to note that the number of studies that support the validity of hypothesis is more than that that support the invalidity of EKC hypothesis for both single-country and panel-country studies.

SINGLE-COUNTRY STUDIES

This chapter starts by discussing single-country studies. Roca, Padilla, Farre, and Galletto (2001) examined the relationship between economic growth and atmospheric pollution by using Spain's indicators data period from 1980 to 1996. This study that applied Seemingly Unrelated Regression (SUR) technique has proved that the EKC hypothesis is invalid for other pollution indicators (CO_2, NO_x, CH_4, N_2O, and NMVOC) outside SO_2. Similar results obtained by Kunnas and Myllyntaus (2007), which is a similar study that applied according to the subelements. The paper revealed that EKC hypothesis is valid for only SO_2 emissions in Finland for the period from 1800 to 2003. Lindmark (2002) investigated the nexus between CO_2 emissions, technology, fuel prices, and growth in Sweden for historic data period 1870–1997, using structural time series models and reached that the EKC hypothesis is valid only for the period 1920–60. Başar and Temurlenk (2007) tested the hypothesis for Turkey 1950–2000 data period. This research that used carbon emissions as a pollution indicator implied that the EKC hypothesis is invalid in Turkey. Soytaş, Sari, and Ewing (2007) investigated the relationship between energy consumption, income, and carbon emissions for United States 1960–2004 data period, using Toda-Yamamoto causality test. They found that income is not cause carbon emissions in the long run, energy consumption does. Contrary to the inverted U-shaped relationship on which the EKC hypothesis is based, Ang (2008) found a positive relationship between pollution and income in Malaysia. Thus, Ang (2008) implied that the EKC hypothesis is invalid as a result of Johansen cointegration test and Error Correction Model (ECM)-based causality analysis. Another study that argues that the EKC hypothesis is not valid belongs to Akbostancı, Türüt-Aşık, and Tunç (2009). They reached this finding, using Johansen cointegration test with Turkey's 1968–2003 data. He and Richard (2010) tested the same relationship by using 1948–2004 data and found that there is a little evidence in favor of the EKC.

Esteve and Tamarit (2012a,b) studied on the EKC hypothesis for Spain during the period 1857–2007. According to their results based on Stock-Watson-Shin and Arai-Kurozumi-Kejriwal cointegration tests, there is not inverted U-shaped relationship between per capita CO_2 and per capita income. Whereas in another study, they used the same country and same period data but changed the method of analysis (threshold cointegration test) and reached the conclusion that the EKC is valid that mean inverted U-shaped relationship between per capita CO_2 and per capita income. In more recent studies, Autoregressive Distributed Lag (ARDL) analysis has been extensively used to test the EKC hypothesis because of the advantages it provides methodically. One of the studies that applied ARDL analysis and showed that EKC is invalid belongs to Saboori, Sulaiman, and Mohd (2012a,b). They examined the relationship between carbon emissions and economic growth in Indonesia data period

Environmental Kuznets Curve (EKC). http://doi.org/10.1016/B978-0-12-816797-7.00004-7

TABLE 4.1
Single-Country and Multiple-Country Studies on the Validity of the Environmental Kuznets Curve Hypothesis.

Study	Country	Period	Method
SINGLE-COUNTRY STUDIES			
Lindmark (2002)	Sweden	1870–1997	STRM
Kunnas and Myllyntaus (2007)	Finland	1800–2003	Regression
Jalil and Mahmud (2009)	China	1975–2005	ARDL
Fodha and Zaghdoud (2010)	Tunisia	1961–2004	Johansen cointegration test
Iwata et al. (2010)	France	1960–2003	ARDL
Nasir and Rehman (2011)	Pakistan	1972–2008	Johansen cointegration test
Esteve and Tamarit (2012a,b)	Spain	1857–2007	Threshold cointegration test
Fosten et al. (2012)	UK	1830–1998	Nonlinear threshold cointegration test and error correction model
Franklin and Ruth (2012)	USA	1800–2000	OLS, Prais-Winsten regression
Saboori et al. (2012a,b)	Malaysia	1980–2009	ARDL
Shahbaz et al. (2012)	Pakistan	1971–2009	ARDL, Gregory-Hansen cointegration test, Granger causality test
Ahmed and Qazi (2013)	Mongolia	1980–2010	Johansen cointegration test
Saboori and Sulaiman (2013)	Malaysia	1980–2009	ARDL, Johansen-Juselius maximum likelihood approach
Shahbaz, Mutascu et al. (2013), Shahbaz, Ozturk et al. (2013)	Romania	1980–2010	ARDL
Shahbaz, Mutascu et al. (2013), Shahbaz, Ozturk et al. (2013)	Turkey	1970–2010	VECM Granger causality
Tiwari et al. (2013)	India	1966–2011	ARDL, VECM Granger causality
Lau et al. (2014)	Malaysia	1970–2008	ARDL, Granger causality
Onafowora and Owoye (2014)	8 countries	1970–2010	ARDL
Shahbaz et al. (2014)	Tunisia	1971–2010	ARDL, VECM Granger causality test
Balaguer and Cantavella (2015)	Spain	1874–2011	ARDL
Bozkurt and Okumus (2015)	Turkey	1966–2011	Hatemi-J cointegration test
Boluk and Mert (2015)	Turkey	1961–2010	ARDL
Al-Mulali et al. (2015a)	Kenya	1980–2012	ARDL
Tutulmaz (2015)	Turkey	1968–2007	Johansen and Engle-Granger cointegration tests
Abdulrazag and Alrajhi (2016)	Saudi Arabia	1971–2013	ARDL, VECM Granger causality
Alam et al. (2016)	Brazil, China, India, Indonesia	1970–2012	ARDL
Ali et al. (2016)	Malaysia	1971–2012	ARDL, DOLS, Granger causality test
Aslan and Gözbaşı (2016)	China	1977–2013	FMOLS, pairwise Granger causality test
Javid and Sharif (2016)	Pakistan	1972–2013	ARDL, VECM cointegration test

TABLE 4.1
Single-Country and Multiple-Country Studies on the Validity of the Environmental Kuznets Curve Hypothesis.—cont'd

Study	Country	Period	Method
Kılıç and Akalın (2016)	Turkey	1960–2011	ARDL
Lebe (2016)	Turkey	1960–2010	ARDL, Granger causality test
Sugiawan and Managi (2016)	Indonesia	1971–2010	ARDL
Waluyo and Terawaki (2016)	Indonesia	1962–2007	ARDL
Zambrano-Monserrate et al. (2016b)	Brazil	1971–2011	ARDL, Granger causality test
Zambrano-Monserrate et al. (2016c)	Iceland	1960–2010	ARDL
Zambrano-Monserrate et al. (2016a)	Ecuador	1971–2011	ARDL, Granger causality test
Rafindadi (2016)	Japan	1961–2012	ARDL
Vita et al. (2016)	Turkey	1960–2009	DOLS
Ahmad et al. (2017)	Croatia	1992Q1 –2011Q1	ARDL, VECM Granger causality, DOLS, FMOLS
Ali et al. (2017)	Pakistan	1960–90	Johansen cointegration test
Och (2017)	Mongolia	1981–2012	ARDL, VECM Granger causality test
Aslan et al. (2018)	USA	1966–2013	Bootstrap rolling window estimation method
Balaguer and Cantavella (2018)	Australia	1950–2014	ARDL
Ravanoğlu et al. (2018)	Kyrgyzstan	1990–2013	ARDL
PANEL-COUNTRY STUDIES			
Markandya et al. (2006)	EU	1870–2001	Panel FE, panel RE
Apergis and Payne (2009)	Central America countries	1971–2004	Pedroni cointegration, FMOLS
Atici (2009)	Central and Eastern Europe	1980–2002	OLS with FE and RE
Lean and Smyth (2010)	ASEAN	1980–2006	Johansen cointegration, panel DOLS
Pao and Tsai (2011)	BRIC countries	1992–2007	OLS
Arouri et al. (2012)	MENA	1981–2005	LM cointegration, mean-group estimation
Donfouet et al. (2013)	EU	19611–2009	Panel GMM and spatial panel
Farhani and Shahbaz (2014)	MENA	1980–2009	Pedroni cointegration, FMOLS, DOLS
Shafiei and Salim (2014)	OECD countries	1980–2011	Johansen cointegration, Westerlund cointegration, GMM, AMG
Heidari, Katircioglu, and Saeidpour (2015)	ASEAN	1980–2008	PSTR model
Al-Mulali, Weng-Wai, Sheau-Ting, and Mohammed (2015c)	Panel of higher-income countries	1980–2008	OLS with FE, GMM
Kasman and Duman (2015)	EU countries	1992–2010	Pedroni cointegration, FMOLS

Continued

TABLE 4.1

Single-Country and Multiple-Country Studies on the Validity of the Environmental Kuznets Curve Hypothesis.—cont'd

Study	Country	Period	Method
Jebli et al. (2016)	OECD countries	1980–2010	Pedroni cointegration, panel FMOLS, panel DOLS
Dogan and Seker (2016)	EU	1980–2012	Panel DOLS
Bilgili et al. (2016)	OECD countries	1977–2010	Pedroni cointegration, FMOLS, DOLS
Al-Mulali and Ozturk (2016)	Panel of 27 advanced countries	1990–2012	Kao cointegration, Johansen cointegration, FMOLS
Ahmad et al. (2016)	EU	1980–2010	Panel ARDL
Charfeddine and Mrabet (2017)	MENA	1995–2007	Panel DOLS and panel FMOLS
Pablo-Romero et al. (2017)	EU	1995–2009	OLS with RE
Rauf et al. (2018)	Belt and Road Initiative countries	1981–2016	Mean-group estimator
Dong et al. (2018)	Asia-Pacific countries	1970–2016	Panel FMOLS and mean-group estimator
Ulucak and Bilgili (2018)	45 countries	1961–2013	CUP-FM, CUP-BC
Hu et al. (2018)	Developing countries	1996–2012	Panel DOLS and panel FMOLS

AMG, augmented mean group estimation; *ARDL*, autoregressive distributed lag; *CUP-BC*, Continuously Updated Bias Corrected; *CUP-FM*, Continuously Updated Fully Modified; *DOLS*, dynamic ordinary least squares; *GMM*, generalized method of moments; *MENA*, middle east and north africa; *OLS*, ordinary least squares; *PSTR*, panel smooth transition regression model; *STRM*, structural time series model; *VECM*, vector error correction model.

from 1971 to 2007, including energy consumption and foreign trade. Using the same method, they obtained an opposite result for the EKC from their analysis with the Malaysia' 1980–2009 data. But unlike the analysis done for Indonesia in this study, energy consumption and foreign trade data were excluded from the econometric model. In addition, a causality test performed using the vector error correction model following the boundary test. Saboori and Sulaiman (2013) examined the nexus between environmental degradation, economic growth, and energy consumption for Malaysia and proved that the validity of the EKC depends on energy consumption. They suggested that decreasing energy consumption appears to be an effective way to control carbon emissions but simultaneously hinder economic growth. Koçak (2014) is another study that tests the EKC hypothesis with ARDL approach and proved that this hypothesis is invalid in Turkey for data period from 1960 to 2010.

Robalino-Lopez, Garcia-Ramos, Golpe, and Mena-Nieto (2014) analyzed whether the EKC hypothesis is held within the period 1980–2025 in Ecuador, using cointegration tests based on Stock and Watson (2010) and reached that the hypothesis is invalid. Boopen

and Vinesh (2015) investigated the relationship among GDP and CO_2 in Mauritania for 1975–2009 data period. They used Johansen cointegration test, and they were unable to prove the existence of a reasonable turning point and thus no EKC inverted-U shape was obtained. The same relation was investigated by Erdoğan, Türköz, and Görüş (2015) for Turkey. There was no evidence of validity of the EKC in the study covering the 1975–2010 periods, using ARDL boundary test and Toda-Yamamoto causality analysis. Jebli and Youssef (2015) preferred the vector error correction model for causality test despite applied ARDL analysis in their study for Tunisia. Their analysis included that economic growth and carbon emissions as well as trade, renewable, and nonrenewable energy consumption variables. Their results suggested that the inverted U-shaped EKC hypothesis is not supported graphically and analytically in the long run. Robalino-Lopez, Mena-Nieto, Garcia-Ramos, and Golpe (2015) studied in detail how changes in the driving forces of the economy affect carbon emissions for Venezuela 1980 2025 data period. Their analysis results based on SUR method implied that Venezuela does not fulfill the EKC hypothesis, but the country could be on the way

TABLE 4.2
Single-Country and Multiple-Country Studies on the Invalidity of the Environmental Kuznets Curve Hypothesis.

Study	Country	Period	Method
SINGLE-COUNTRY STUDIES			
Roca et al. (2001)	Spain	1980−96	SUR
Başar and Temurlenk (2007)	Turkey	1950−2000	Regression
Soytaş et al. (2007)	USA	1960−2004	Toda-Yamamoto causality test
Ang (2008)	Malaysia	1971−99	Johansen cointegration, ECM-based causality test
Akbostancı et al. (2009)	Turkey	1968−2003	Johansen cointegration test
He and Richard (2010)	Canada	1948−2004	PLR model and Hamilton's model
Esteve and Tamarit (2012a,b)	Spain	1857−2007	Stock-Watson-Shin and Arai-Kurozumi-Kejriwal cointegration test
Saboori et al. (2012a,b)	Indonesia	1971−2007	ARDL
Koçak (2014)	Turkey	1960−2010	ARDL
Robalino-Lopez et al. (2014)	Ecuador	1980−2025	Cointegration test
Boopen and Vinesh (2015)	Mauritania	1975−2009	Johansen cointegration test, VAR, OLS
Erdoğan et al. (2015)	Turkey	1975−2010	ARDL, Toda-Yamamoto Granger causality
Al-Mulali et al. (2015b)	Vietnam	1981−2011	ARDL
Ozturk and Al-Mulali (2015)	Cambodia	1996−2012	GMM, 2SLS
Robalino-Lopez et al. (2015)	Venezuela	1980−2025	SUR
Jebli and Youssef (2015)	Tunisia	1980−2009	ARDL, VECM Granger causality test
Erdoğan et al. (2015)	Turkey	1975−2010	ARDL, Toda-Yamamoto Granger causality
Al-Mulali et al. (2015b)	Vietnam	1981−2011	ARDL
Ozturk and Al-Mulali (2015)	Cambodia	1996−2012	GMM, 2SLS
Robalino-Lopez et al. (2015)	Venezuela	1980−2025	SUR
PANEL-COUNTRY STUDIES			
Ozcan (2013)	Middle East countries	1990−2008	Westerlund cointegration, Panel FMOLS
Boluk and Mert (2014)	EU countries	1990−2008	OLS with FE
Lopez-Menendez, Perez, and Moreno (2014)	EU countries	1996−2010	OLS with FE and RE
Ajmi, Hammoudeh, Nguyen, and Sato (2015)	G7 countries	1960−2010	Ng-Perron unit root, causality test
Al-Mulali et al. (2016)	58 countries	1980−2009	Panel FE and panel GMM
Dogan et al. (2017)	OECD	1995−2010	LM cointegration, DOLS
Abid (2017)	EU	1990−2011	Panel GMM

PLR, Partially Linear Regression.

to achieve environmental stabilization in the medium run. Al-Mulali, Saboori, and Ozturk (2015b) reached the findings that the EKC is invalid for the case of Vietnam, using ARDL analysis. Although they used the same data period and the same method, they found the opposite result when they study for Kenya. Ozturk and Al-Mulali (2015) investigated whether better governess and corruption control help to form the inverted U-shaped relationship between income and pollution in Cambodia for the period of 1996–2012. The results from Generalized Method of Moments (GMM) and the Two-Stage Least Squares (2SLS) revealed that GDP, urbanization, energy consumption, and trade openness increase CO_2 emission, whereas the control of corruption and governess reduces carbon emission and ultimately showed that the EKC hypothesis is invalid for Cambodia. Yurttagüler and Kutlu (2017) tested the EKC hypothesis in Turkey, using the data for 1960–2011. Their results indicated that there is an N-shaped relationship between income and carbon emission.

Although there are numerous time-series studies advocating that this hypothesis is invalid in studies conducted to test the EKC hypothesis, a relatively large number of studies have proved that this hypothesis is valid. Jalil and Mahmud (2009) examined the long-run relationship between carbon emissions, energy consumption, income, and foreign trade in the case of China by using time-series data of 1975–2005. They aimed testing the EKC hypothesis for CO_2 emission and per capita real GDP by using ARDL method and Granger causality analysis. Nasir and Rehman (2011) also researched for Pakistan using the same indicators, and the results again confirmed the validity of EKC hypothesis for the period 1972–2008. The analysis results indicated that the EKC hypothesis is valid; in addition, carbon emissions are mainly determined by income and energy consumption. Another study that analyzes for Pakistan using carbon emissions, energy consumption, growth, and trade indicators is the research of Shahbaz, Lean, and Shabbir (2012). Their analyses that are based on ARDL, Gredory-Hansen cointegration, and Granger causality tests proved that existence of EKC hypothesis for the period from 1971 to 2009. Shahbaz, Mutascu, and Azim (2013), Shahbaz, Ozturk, Afza, and Ali (2013) tested the EKC hypothesis for Romania this time, not including of trade variable in analysis. In this study, they applied ARDL boundary test and revealed that the EKC is valid for the period of 1980–2010. They analyzed this relationship in the same year for Turkey and confirmed the presence of the EKC hypothesis. Ahmed and Qazi (2013) tested the EKC hypothesis for CO_2 in Mongolia data period from 1980 to 2010, using Johansen cointegration test with major explanatory variables as energy consumption, economic growth, and trade openness. Their results confirmed the existence EKC hypothesis among the variables both in long run and short run except trade

openness. Tiwari, Shahbaz, and Hye (2013) also examined the case of India with the same indicators, using ARDL analysis and Vector Error Coreection Model (VECM) Granger causality test. They revealed that existence of the EKC hypothesis, and coal consumption and trade openness contributes to CO_2 emission.

Fodha and Zaghdoud (2010) investigated this hypothesis for Tunisia data period from 1961 to 2004 by using Johansen cointegration test. They used CO_2 and SO_2 emissions as the environmental pollution indicators and the GDP as an economic indicator. Their results proved that there is an inverted U-shaped relationship between SO_2 emissions and GDP, with income turning point approximately equals to $1200 but was found monotonically a relationship between CO_2 and GDP. Another study using the variables of carbon emission and sulfur dioxide emission as pollution indicators belongs to Fosten et al. (2012). They investigated whether the EKC hypothesis is valid for the United Kingdom by using nonlinear threshold cointegration test and error correction model and reached that there is an inverted U-shaped relationship between per capita CO_2 and SO_2 emissions and per capita GDP. Och (2017) used nitrous oxide (NO_x) emissions as an indicator of environmental pollution and analyzed the EKC hypothesis in Mongolia for the 1981–2012 period, using ARDL and VECM Granger causality tests. A highly significant and robust long-run U-shaped relationship between NO_x emissions and income was found, meaning that the existence of the EKC hypothesis.

Iwata, Okada, and Samreth (2010) studied for France by considering the role of nuclear energy, and their ARDL estimation results provided evidence supporting the EKC hypothesis for the 1960–2003 data period. Franklin and Ruth (2012) conducted a regression analysis using the 1800–2000 period data from the United States and found the inverted U-shaped relationship between pollution and growth. Lau, Choong, and Eng (2014) provided a different contribution to the literature by including foreign direct investment in analysis for Malaysia. They attempted to examine the EKC hypothesis for Malaysia in the presence of foreign direct investment and trade openness both in the short and long run for the period from 1970 to 2008. Their analysis results indicated the inverted U-shaped relationship between economic growth and CO_2 emission. Ali, Abdullah, and Azam (2016) investigated the impact of foreign direct investment as well as financial development, energy consumption, and trade openness on carbon emission for Malaysia, using ARDL boundary test, DOLS method, and Granger causality analysis; they reached same results with Lau et al. (2014). Javid and Sharif (2016) analyzed the effects of financial development, income, energy consumption, and trade openness on CO_2 emissions in Pakistan for the period from 1972 to 2013. In this study, where ARDL analysis was the basic research methodology, the validity of the hypothesis was revealed. Lebe (2016) achieved the same result for Turkey, using same indicators and same analysis technique.

Onafowora and Owoye (2014) investigated the EKC hypothesis for each of eight different countries (Brazil, China, Egypt, Japan, Mexico, Nigeria, South Korea, and South Africa) with ARDL analysis, using economic growth, energy consumption, population density, trade openness, and CO_2 emission indicators. The estimate results showed that the inverted U-shaped EKC hypothesis holds in Japan and South Korea. Bozkurt and Okumus (2015) considered energy consumption, population density, and trade openness in testing the hypothesis such as Onafowora and Owoye (2014). In this study that applied Hatemi-J cointegration test with 1966−2011 data is proved the validity of the hypothesis for Turkey. A study to test the EKC hypothesis with same indicators was also conducted by Abdulrazag and Alrajhi (2016) for Saudi Arabia. Results of ARDL analysis and VECM causality test with 1971−2013 data revealed the validity of the hypothesis. Alam, Murad, Noman, and Ozturk (2016) examined the impacts of income, energy consumption, and population growth on CO_2 emissions in Brazil, China, India, and Indonesia for the 1970−2012 data period. Their results suggested that the EKC hypothesis is valid only in Brazil, China, and Indonesia.

Shahbaz, Khraief, Uddin, and Ozturk (2014) investigated the existence of the EKC hypothesis in Tunisia for the period of 1971−2010. They applied ARDL boundary test and causality analysis based on VECM and proved the presence of this hypothesis. Balaguer and Cantavella (2015) included fuel oil prices in analysis while testing the EKC hypothesis for Spain. ARDL test results supported the hypothesis. Boluk and Mert (2015) tried to demonstrate the reducing effect of renewable energy sources on greenhouse gas emission, using ARDL approach for Turkey; their results suggested an inverted U-shaped relationship between per capita green gas emission and income. Tutulmaz (2015) using Johansen and Engle-Granger cointegration tests reached the first phases of an inverted U-shaped EKC relationship in his study for Turkey data period from 1968 to 2007. Aslan and Gözbaşı (2016) investigated the validity of the EKC hypothesis for the period among 1977 and 2013 by using the subelements of CO_2 emissions

in China. Fully Modified Ordinary Least Squares (FMOLS), Pairwise Granger causality test results suggested that the EKC hypothesis is valid for CO_2 emissions from gaseous fuel consumption, liquid fuel consumption, solid fuel consumption, and transportation but invalid for aggregate CO_2 emissions: CO_2 emissions from residential buildings and commercial and public services, from electricity and heat production, and from manufacturing industries and constructions. Kılıç and Akalın (2016, pp. 49−60) used per capita CO_2 emission as the indicator of environmental pollution and per capita income as the indicator of economic growth. Their results obtained from ARDL boundary analysis show that there is an inverted U-shaped relationship between the variables for 1960−2011 periods.

Sugiawan and Managi (2016) explored the existence of EKC hypothesis in Indonesia by using the analysis and reached that the EKC hypothesis is valid (turning point is 7729 USD per capita). Waluyo and Terawaki (2016) also made an ARDL analysis for Indonesia but also considered deforestation. Test results supported the long-run inverted-U relationship, thus, while the deforestation rate increases at the initial stage of economic growth, it declined after a threshold point (990.4 USD). Zambrano-Monserrate, Garcia-Alban, and Henk-Vera (2016) conducted different studies for Brazil, Iceland, and Ecuador and applied ARDL analysis in each study. The results of three studies revealed the validity of the EKC hypothesis. Rafindadi (2016) used ARDL analysis to test the presence of EKC in Japan for the period from 1961 to 2012. This analysis of the period covering energy disaster and deteriorating income proved that the EKC is valid despite deteriorating income. Vita, Katircioglu, Altinay, Fethi, and Mercan (2016) while testing the EKC hypothesis for Turkey took the subject from a different perspective in the context of tourism development. They found that tourist arrivals, growth, and energy consumption exert a positive and statistically significant impact on CO_2 emissions in the long run. In addition, empirical results supported the existence of EKC hypothesis. Ahmad et al. (2017) analyzed the relationship between CO_2 emissions and economic growth in Croatia for the 1992−2011 periods, using ARDL analysis, VECM Granger causality test, Dynamic Ordinary Least Squares (DOLS), and FMOLS estimation methods. Their results confirmed the EKC hypothesis. Aslan, Destek, and Okumus (2018) investigated the hypothesis in the United States for the period from 1966 to 2013. They applied bootstrap rolling window estimation method and reached that the hypothesis is valid. Balaguer and

Cantavella (2018) examined the role of education on EKC in Australia. According to this study, the EKC hypothesis is valid; in addition, the effect of education is positive on the U-shaped relationship. Ravanoğlu et al. (2018) investigated the validity of EKC hypothesis in Kyrgyzstan, using ARDL analysis. They findings suggested that the EKC hypothesis is valid in the long run. Finally, a different study conducted by Ali, Ashraf, Bashir, and Cui (2017) explored the EKC hypothesis in relation to Green Revolution in Pakistan, applying Johansen cointegration test. They reached that the EKC hypothesis in relation to the Green Revolution is not valid.

MULTIPLE-COUNTRY STUDIES

Panel-country studies are henceforth discussed. Lean and Smyth (2010) analyzed the relationship between CO_2 emissions and economic growth for the panel of Association of Southeast Asian Nations (ASEAN) countries for the 1980−2006 periods, using the Johansen cointegration test and DOLS. Their results confirmed the EKC hypothesis. Moreover, Heidari, Katircioğlu, and Saeidpour (2015) also supported the validity of the EKC hypothesis for the ASEAN countries. Kasman and Duman (2015) investigated the effects of energy consumption, the real income, trade, and urbanization on the level of emissions for the EU countries over the period 1992−2010 by using the FMOLS. They reached that the EKC hypothesis is valid. The existence of the EKC hypothesis was supported for Central American countries in the study of Apergis and Payne (2009). Jebli, Youssef, and Ozturk (2016) suggested that the EKC hypothesis is valid for the Organization for Economic Cooperation and Development (OECD) countries by applying the FMOLS and the DOLS on renewable energy, nonrenewable energy consumption, economic growth, trade, and CO_2 emissions for 1980−2010 periods. Bilgili, Kocak, and Bulut (2016) and Shafiei and Salim (2014) reached the same conclusion for OECD members. On the contrary, Dogan, Seker, and Bulbul (2017) found the evidence of invalidity of the EKC hypothesis for OECD countries by using the DOLS for 1995−2010 periods.

Boluk and Mert (2014) studied on EKC hypothesis for the European Union (EU) members during the period 1990−2008. According to their results based on the Ordinary Least Squares (OLS) with Random Effects (RE), there is not inverted U-shaped relationship between CO_2 emissions and income. In a similar study, Lopez-Menendez, Perez, and Moreno (2014) also reported the invalidity of the EKC hypothesis for the EU

countries by using the data on CO_2 emissions, income, the square of income, the cube of income, and renewable energy for the period of 1996−2010, whereas in another study, Dogan and Seker (2016) used the same country but different period data (1980−2012) and different method of analysis (the panel DOLS). They reached the conclusion that EKC is valid; in another mean, an inverted U-shaped relationship between CO_2 and income. In a more recent study by Pablo-Romero (2017), which applied the OLS with RE on the data from 1995 to 2009 showed that EKC is valid for the EU countries.

There are very recent studies that cover different panel of countries use different methodologies. Rauf et al. (2018) analyzed the effects of income per capita, square of income per capita, population growth, and energy consumption on environmental degradation measures (deforestation, water quality, and air pollution) for the Belt and Initiative countries over the period 1981−2016 by using the Westerlund cointegration test and mean group estimators. Results showed the validity of the EKC hypothesis. Dong, Sun, Li, and Liao (2018) examined the energy-growth-environment nexus by focusing on relationship between natural gas consumption, CO_2 emissions and GDP per capita for a panel of 14 Asia-Pacific economies for the years 1970−2016. Empirical results revealed that natural gas consumption increased the environmental degradation, and the reverted U-shaped relationship is found between carbon emissions and income per capita, implying the validity of the EKC hypothesis. Referring to Hu, Xie, Fang, and Zhang (2018), the EKC was verified in a panel of 25 major developing countries for the years 1996−2012 by using the FMOLS and the DOLS estimators.

As a summary of a review of the EKC hypothesis for panel-country studies, the presence of EKC hypothesis is confirmed for the EU countries in Markandya et al. (2006), Donfuet et al. (2013), Kasman and Duman (2015), Dogan and Seker (2016), Ahmad et al. (2016), and Pablo-Romero et al. (2017); for the OECD countries in Shafiei and Salim (2014), Jebli et al. (2016), Bilgili et al. (2016), and Dogan et al. (2017); for the Middle East and North Africa (MENA) economies in Arouri, Youssef, M'henni, and Rault (2012), Farhani and Shahbaz (2014), and Charfeddine and Mrabet (2017); for the ASEAN countries in Lean and Smyth (2010) and Heidari et al. (2015); for some other panel of countries in Apergis and Payne (2009), Pao and Tsai (2011), Al-Mulali, Solarin, and Ozturk (2015a), Al-Mulali and Ozturk (2016), Rauf et al. (2018), and Ulucak and Bilgili (2018).

CONCLUSION

In the late 1990s, there was an explosion in the empirical literature investigating the relationship between economic growth and environmental degradation. Much of these studies' main purpose is to test the EKC hypothesis, which is based on the idea that there is a positive relationship between the two variables in the early stages of economic development and that it is transformed into a negative relationship in the later stages of development. This chapter has focused on the empirical literature on the EKC hypothesis. The studies in the literature are divided into two groups as single-country analysis and panel-country analysis. In addition, all studies are classified according to the results obtained in terms of whether EKC hypothesis is valid or not.

As a result of this research, firstly, it is worth to note that the number of studies that support the validity of hypothesis is more than that that support the invalidity of EKC hypothesis for both single-country and panel-country studies. Another result is that the number of single-country studies is greater than that of panel-country studies. This result can be explained by the fact that the determination of the EKC relationship and the threshold income level in a single country makes it easier to direct national policies.

The research has shown that the literature on the EKC hypothesis is still up to date although it has a very old history. In particular, econometric developments have allowed this hypothesis to be tested by different methods. Furthermore, developments in economic science require the continuous updating of the models used to test this hypothesis. This dynamic process explains the popularity of the literature on the EKC hypothesis and the differences in the results achieved. Different results from the empirical studies have shown that there is no consensus on the validity of the hypothesis. Therefore, the literature on the hypothesis has been extensively examined, and a contribution has been made to determine the starting point of future studies.

REFERENCES

Abdulrazag, B., & Alrajhi, A. (2016). The validity of environmental Kuznets curve hypotheis in the kingdom of Saudi Arabia: ARDL bounds testing approach to cointegration. *International Review of Management and Business Research*, 5(4), 1450−1464.

Abid, M. (2017). Does economic, financial and institutional developments matter for environmental quality? A comparative analysis of EU and MEA countries. *Journal of Environmental Management, 188, 183−194.*

Ahmad, N., Du, L., Lu, J., Wang, J., Li, H., & Hashmi, M. Z (2017). Modelling the CO2 emissions and economic growth in Croatia. *Is there any environmental Kuznets curve? Energy, 123,* 164−172.

Ahmad, A., Zhao, Y., Shahbaz, M., Bano, S., Zhang, Z., Wang, S., & Liu, Y. (2016). Carbon emissions, energy consumption and economic growth: An aggregate and disaggregate analysis of the Indian economy. *Energy Policy, 96,* 131−143.

Ahmed, K., & Qazi, A. Q. (2013). Environmental Kuznets curve for CO_2 emission in Mongolia: An empirical analysis. *Management of Environmental Quality: An International Journal, 25*(4), 505−516.

Ajmi, A. N., Hammoudeh, S., Nguyen, D. K., & Sato, J. R. (2015). On the relationships between CO_2 emissions, energy consumption and income: The importance of time variation. *Energy Economics, 49,* 629−638.

Akbostancı, E., Türüt-Aşık, S., & Tunç, G.İ. (2009). The relationship between income and environment in Turkey: Is there an environmental Kuznets curve? *Energy Policy, 37,* 861−867.

Al-Mulali, U., & Ozturk, I. (2016). The investigation of environmental Kuznets curve hypothesis in the advanced economies: The role of energy prices. *Renewable and Sustainable Energy Reviews, 54,* 1622−1631.

Al-Mulali, U., Saboori, B., & Ozturk, I. (2015b). Investing the environmental Kuznets curve hypothesis in Vietnam. *Energy Policy, 76,* 123−131.

Al-Mulali, U., Solarin, S. A., & Ozturk, I. (2015a). Investigating the presence of the environmental Kuznets curve (EKC) hypothesis in Kenya: An autoregressive distributed lag (ARDL) approach. *Natural Hazards, 80*(3), 1729−1747.

Al-Mulali, U., Weng-Wai, C., Sheau-Ting, L., & Mohammed, A. H. (2015c). Investigating the environmental Kuznets curve (EKC) hypothesis by utilizing the ecological footprint as an indicator of environmental degradation. *Ecological Indicators, 48,* 315−323.

Al-Mulali, U., Solarin, S. A., Sheau-Ting, L., & Ozturk, I. (2016). Does moving towards renewable energy cause water and land inefficiency? An empirical investigation. *Energy Policy, 93,* 303−314.

Alam, M., Murad, W., Noman, A. H., & Ozturk, I. (2016). Relationship among carbon emissions, economic growth, energy consumption and population growth: Testing environmental Kuznets curve hypothesis for Brazil, China, India and Indonesia. *Ecological Indicators, 70,* 466−479.

Ali, W., Abdullah, A., & Azam, M. (2016). Re-visiting the environmental Kuznets curve hypothesis for Malaysia: Fresh evidence from ARDL bounds testing approach. *Renewable and Sustainable Energy Reviews, 77,* 990−1000.

Ali, G., Ashraf, A., Bashir, M. K., & Cui, S. (2017). Exploring environmental Kuznets curve (EKC) in relation to green revolution: A case study of Pakistan. *Environmental Science and Policy, 77,* 166−171.

Ang, J. B. (2008). Economic development, pollutant emissions and energy consumption in Malaysia. *Journal of Policy Modeling, 30,* 271−278.

Apergis, N., & Payne, J. E. (2009). CO_2 emissions, energy usage, and output in Central America. *Energy Policy, 37*(8), 3282−3286.

Arouri, M. E. H., Youssef, A. B., M'henni, H., & Rault, C. (2012). Energy consumption, economic growth and CO_2 emissions in Middle East and North African countries. *Energy Policy*, 45, 342−349.

Aslan, A., Destek, M. A., & Okumus, I. (2018). Bootstrap rolling window estimation approach to analysis of the environment Kuznets curve hypothesis: Evidence from the USA. *Environmental Science and Pollution Research*, 25(3), 2402−2408.

Aslan, A., & Gözbaşı, O. (2016). Environmental Kuznets curve hypothesis for sub-elements of the carbon emissions in China. *Natural Hazards*, 82, 1327−1340.

Atici, C. (2009). Carbon emissions in Central and Eastern Europe. Environmental Kuznets curve and implications for sustainable development. *Sustainable Development*, 17(3), 155−160.

Balaguer, J., & Cantavella, M. (2015). Estimating the environmental Kuznets curve for Spain by considering fuel oil prices (1874−2011). *Ecological Indicators*, 60, 853−859.

Balaguer, J., & Cantavella, M. (2018). The role of education in the environmental Kuznets Curve. Evidence from Australian data. *Energy Economics*, 70, 289−296.

Başar, S., & Temurlenk, M. S. (2007). Çevreye Uyarlanmış Kuznets Eğrisi: Türkiye Üzerine Bir Uygulama. *Atatürk Üniversitesi İktisadi ve İdari Bilimler Dergisi*, 21(1), 1−12.

Bilgili, F., Kocak, E., & Bulut, U. (2016). The dynamic impact of renewable energy consumption on CO_2 emissions: A revisited environmental Kuznets curve approach. *Renewable and Sustainable Energy Reviews*, 54, 838−845.

Boluk, G., & Mert, M. (2014). Fossil & renewable energy consumption, GHGs (greenhouse gases) and economic growth: Evidence from a panel of EU (European Union) countries. *Energy*, 74, 439−446.

Boluk, G., & Mert, M. (2015). The renewable energy, growth and environmental Kuznets curve in Turkey: An ARDL approach. *Renewable and Sustainable Energy Reviews*, 52, 587−595.

Boopen, S., & Vinesh, S. (2015). On the relationship between CO2 emissions and economic growth: The Mauritian experience. In *CSAE 25th anniversary conference 2011: Economic development in Africa 2011* (pp. 1−25).

Bozkurt, C., & Okumus, İ. (2015). The effects of economic growth, energy consumption, trade openness and population density on CO_2 emissions in Turkey: A cointegration analysis with structural breaks. *Mustafa Kemal University Journal of Social Sciences Institute*, 12(32), 23−35.

Charfeddine, L. & Mrabet, Z., The impact of economic development and social-political factors on ecological footprint: A panel data analysis for 15 MENA countries. Renewable and Sustainable Energy Reviews, 76,138−154.

Dogan, E., & Seker, F. (2016). Determinants of CO_2 emissions in the European Union: The role of renewable and non-renewable energy. *Renewable Energy*, 94, 429−439.

Dogan, E., Seker, F., & Bulbul, S. (2017). Investigating the impacts of energy consumption, real GDP, tourism and trade on CO_2 emissions by accounting for cross-sectional dependence: A panel study of OECD countries. *Current Issues in Tourism*, 20(16), 1701−1719.

Donfuet, H. P. P., Jeanity, P. W., & Malin, E. A Spatial Dynamic Panel Analysis of the Environmental Kuznets Curve in European Countries. *Discussion paper, Center for Research in Economics and Management (CREM)*, University of Rennes 1, University of Caen and CNRS.

Dong, K., Sun, R., Li, H., & Liao, H. (2018). Does natural gas consumption mitigate CO_2 emissions: Testing the environmental Kuznets curve hypothesis for 14 Asia-Pacific countries. *Renewable and Sustainable Energy Reviews*, 94, 419−429.

Erdoğan, İ., Türköz, K., & Görüş, M.Ş. (2015). Validity of environmental Kuznets curve hypothesis for the Turkish economy. *Dumlupınar University Journal of Social Sciences*, 44, 113−123.

Esteve, V., & Tamarit, C. (2012a). Is there an environmental Kuznets curve for Spain? Fresh evidence from old data. *Economic Modelling*, 29, 2696−2703.

Esteve, V., & Tamarit, C. (2012b). Threshold cointegration and nonlinear adjustment between CO_2 and income: The environmental Kuznets curve in Spain, 1857−2007. *Energy Economics*, 34, 2148−2156.

Farhani, S., & Shahbaz, M. (2014). What role of renewable and non-renewable electricity consumption and output is needed to initially mitigate CO_2 emissions in MENA region? *Renewable and Sustainable Energy Reviews*, 40, 80−90.

Fodha, M., & Zaghdoud, O. (2010). Economic growth and pollutant emissions in Tunisia: An empirical analysis of the environmental Kuznets curve. *Energy Policy*, 38, 1150−1156.

Fosten, J., et al. (2012). Dynamic misspecification in the environmental Kuznets curve: Evidence from CO_2 and SO_2 emissions in the United Kingdom. *Ecological Economics*, 76, 25−33.

Franklin, R. S., & Ruth, M. (2012). Growing up and cleaning up: The environmental Kuznets curve redux. *Applied Geography*, 32, 29−39.

Heidari, H., Katircioğlu, S. T., & Saeidpour, L. (2015). Economic growth, CO_2 emissions, and energy consumption in the five ASEAN countries. *International Journal of Electrical Power & Energy Systems*, 64, 785−791.

He, J., & Richard, P. (2010). Environmental Kuznets curve for CO2 in Canada. *Ecological Economics*, 69, 1083−1093.

Hu, H., Xie, N., Fang, D., & Zhang, X. (2018). The role of renewable energy consumption and commercial services trade in carbon dioxide reduction: Evidence from 25 developing countries. *Applied Energy*, 211, 1229−1244.

Iwata, H., Okada, K., & Samreth, S. (2010). Empirical study on the environmental Kuznets curve for CO_2 in France: The role of nuclear energy. *Energy Policy*, 38, 4057−4063.

Jalil, A., & Mahmud, S. F. (2009). Environment Kuznets curve for CO_2 emissions: A cointegration analysis for China. *Energy Policy*, 37, 5167−5172.

Javid, M., & Sharif, F. (2016). Environmental Kuznets curve and financial development in Pakistan. *Renewable and Sustainable Energy Reviews*, 54, 406−414.

Jebli, M. B., & Youssef, S. B. (2015). The environmental Kuznets curve, economic growth, renewable and non-renewable energy, and trade in Tunisia. *Renewable and Sustainable Energy Reviews, 47*, 173, 18.

Jebli, M. B., Youssef, S. B., & Ozturk, I. (2016). Testing environmental Kuznets curve hypothesis: The role of renewable and non-renewable energy consumption and trade in OECD countries. *Ecological Indicators, 60*, 824–831.

Kasman, A., & Duman, Y. S. (2015). CO_2 emissions, economic growth, energy consumption, trade and urbanization in new EU member and candidate countries: A panel data analysis. *Economic Modelling, 44*, 97–103.

Koçak, E. (2014). Türkiye'de Çevresel Kuznets Eğrisi Hipotezinin Geçerliliği: ARDL Sınır Testi Yaklaşımı. *İşletme ve İktisat Çalışmaları Dergisi, 2*(3), 62–73.

Kunnas, J., & Myllyntaus, T. (2007). The environmental Kuznets curve hypothesis and air pollution in Finland. *Scandinavian Economic History Review, 55*(2), 101–127.

Kılıç, R., & Akalın, G. (2016). *The relationship between environment and economic growth in Turkey: An ARDL bounds testing approach.* Anadolu University Journal of Social Sciences.

Lau, L. S., Choong, C. K., & Eng, Y. K. (2014). Investigation of the environmental Kuznets curve for carbon emissions in Malaysia: Do foreign direct investment and trade matter? *Energy Policy, 68*, 490–497.

Lean, H. H., & Smyth, R. (2010). CO_2 emissions, electricity consumption and output in ASEAN. *Applied Energy, 87*(6), 1858–1864.

Lebe, F. (2016). The environmental Kuznets curve hypothesis: Cointegration and causality analysis for Turkey. *Journal of Donghua University, 17*(2), 177–194.

Lindmark, M. (2002). An EKC-pattern in historical perspective: Carbon dioxide emissions, technology, fuel prices and growth in Sweden 1870–1997. *Ecological Economics, 42*, 333–347.

Lopez-Menendez, A. J., Perez, R., & Moreno, B. (2014). Environmental costs and renewable energy: Re-visiting the environmental Kuznets curve. *Journal of Environmental Management, 145*, 368–373.

Markandya, A., Golub, A., & Pedroso-Galinato, S. (2006). Empirical analysis of national income and SO_2 emissions in selected European Countries. *Environmental & Resource Economics, Springer; European Association of Environmental and Resource Economists, 35*(3), 221–257.

Nasir, M., & Rehman, F. U. (2011). Environmental Kuznets curve for carbon emissions in Pakistan: An empirical investigation. *Energy Policy, 39*, 1857–1864.

Och, M. (2017). Empirical investigation of the environmental Kuznets curve hypothesis for nitrous oxide emissions for Mongolia. *International Journal of Energy Economics and Policy, 7*(1), 117–128.

Onafowora, O. A., & Owoye, O. (2014). Bounds testing approach to analysis of the environment Kuznets curve hypothesis. *Energy Economics, 44*, 47–62.

Ozcan, B. (2013). The nexus between carbon emissions, energy consumption and economic growth in Middle East countries: A panel data analysis. *Energy Policy, 62*, 1138–1147.

Ozturk, I., & Al-Mulali, U. (2015). Investigating the validity of the environmental Kuznets curve hypothesis in Cambodia. *Ecological Indicators, 57*, 324–330.

Pablo-Romero, M. (2017). Residential energy environmental Kuznets curve in the EU-28. *Energy, 125*(C), 44–54.

Pablo-Romero, M. P., Cruz, L., & Barata, E. (2017). Testing the transport energy-environmental Kuznets curve hypothesis in the EU27 countries. *Energy Economics, 62*, 257–269. https://doi.org/10.1016/j.eneco.2017.01.003.

Pao, H. T., & Tsai, C. M. (2011). Multivariate Granger causality between CO_2 emissions, energy consumption, FDI (foreign direct investment) and GDP (gross domestic product): Evidence from a panel of BRIC (Brazil, Russian Federation, India, and China) countries. *Energy, 36*(1), 685–693.

Rafindadi, A. A. (2016). Revisiting the concept of environmental Kuznets curve in period of energy disaster and deteriorating income: Empirical evidence from Japan. *Energy Policy, 94*, 274–284.

Rauf, A., Liu, X., Amin, W., Ozturk, I., Rehman, O. U., & Hafeez, M. (2018). Testing EKC hypothesis with energy and sustainable development challenges: A fresh evidence from belt and road initiative economies. *Environmental Science and Pollution Research*, 1–15.

Ravanoğlu, G. A., et al. (2018). Validity of environmental Kuznets curve hypothesis for Kyrgyzstan economy: ARDL bound test approach. *MANAS Journal of Social Studies, 7*(2), 127–142.

Robalino-Lopez, A., Garcia-Ramos, J. E., Golpe, A. A., & Mena-Nieto, A. (2014). System dynamics modeling and the environmental Kuznets curve in Ecuador (1980–2025). *Energy Policy, 67*, 923–931.

Robalino-Lopez, A., Mena-Nieto, A., Garcia-Ramos, J. E., & Golpe, A. A. (2015). Studying the relationship between economic growth, CO_2 emissions, and the environmental Kuznets curve in Venezuela (1980–2025). *Renewable and Sustainable Energy Reviews, 41*, 602–614.

Roca, J., Padilla, E., Farre, M., & Galletto, V. (2001). Economic growth and atmospheric pollution in Spain: Discussing the environmental Kuznets curve hypothesis. *Ecological Economics, 39*, 85–99.

Saboori, B., & Sulaiman, J. (2013). Environmental degradation, economic growth and energy consumption: Evidence of the environmental Kuznets curve in Malaysia. *Energy Policy, 60*, 892–905.

Saboori, B., Sulaiman, J., & Mohd, S. (2012a). An empirical analysis of the environmental Kuznets curve for CO_2 emissions in Indonesia: The role of energy consumption and foreign trade. *International Journal of Economics and Finance, 4*(2), 243–251.

Saboori, B., Sulaiman, J., & Mohd, S. (2012b). Economic growth and CO_2 emissions in Malaysia: A cointegration analysis of the environmental Kuznets curve. *Energy Policy, 51*, 184–191.

Shafiei, S., & Salim, R. A. (2014). Non-renewable and renewable energy consumption and CO_2 emissions in OECD countries: A comparative analysis. *Energy Policy, 66*, 547–556.

Shahbaz, M., Khraief, N., Uddin, G. Z., & Ozturk, I. (2014). Environmental Kuznets curve in an open economy: A bounds testing and causality analysis for Tunisia. *Renewable and Sustainable Energy Reviews, 34*, 325–336.

Shahbaz, M., Lean, H. H., & Shabbir, M. S. (2012). Environmental Kuznets curve hypothesis in Pakistan: Cointegration and Granger causality. *Renewable and Sustainable Energy Reviews, 16,* 2947−2953.

Shahbaz, M., Mutascu, M., & Azim, P. (2013). Environmental Kuznets curve in Romania and the role of energy consumption. *Renewable and Sustainable Energy Reviews, 18,* 165−173.

Shahbaz, M., Ozturk, I., Afza, T., & Ali, A. (2013). Revisiting the environmental Kuznets curve in a global economy. *Renewable and Sustainable Energy Reviews, 25,* 494−502.

Soytaş, U., Sari, R., & Ewing, B. T. (2007). Energy consumption, income, and carbon emissions in the United States. *Ecological Economics, 62,* 482−489.

Stock, J. H., & Watson, M. V. A simple estimator of cointegrating vectors in higher order integrated systems. *Econometrica, 61,* 783−820.

Sugiawan, Y., & Managi, S. (2016). The environmental Kuznets curve in Indonesia: Exploring the potential of renewable energy. *Energy Policy, 98,* 187−198.

Tiwari, A. K., Shahbaz, M., & Hye, Q. M. A. (2013). The environmental Kuznets curve and the role of coal consumption in India: Cointegration and causality analysis in an open economy. *Renewable and Sustainable Energy Reviews, 18,* 519−527.

Tutulmaz, O. (2015). Environmental Kuznets Curve time series application for Turkey: Why controversial results exist for similar models? *Renewable and Sustainable Energy Reviews, 50,* 73−81.

Ulucak, R., & Bilgili, F. (2018). A reinvestigation of EKC model by ecological footprint measurement for high, middle and low income countries. *Journal of Cleaner Production, 188,* 144−157.

Vita, D., Katircioğlu, S., Altinay, L., Fethi, S., & Mercan, M. (2016). Revisiting the environmental Kuznets curve hypothesis in a tourism development context. *Environmental Science and Pollution Research, 22*(21), 16652−16663.

Waluyo, E. A., & Terawaki, T. (2016). Environmental Kuznets curve for deforestation in Indonesia: An ARDL bounds testing approach. *Journal of Economic Cooperation and Development, 37*(3), 87−108.

Yurttagüler, İ. M., & Kutlu, S. (2017). An econometric analysis of the environmental Kuznets curve: The case of Turkey. *Alphanumeric Journal The Journal of Operations Research, Statistics, Econometrics and Management Information Systems, 5*(1), 115−126.

Zambrano-Monserrate, M. A., Garcia-Alban, F. F., & Henk-Vera, K. A. (2016a). Bounds testing approach to analyze the existence of an environmental Kuznets curve in Ecuador. *International Journal of Energy Economics and Policy, 6*(2), 159−166.

Zambrano-Monserrate, M. A., Troccoly-Quiroz, A., & Pacheco-Borja, M. J. (2016c). Testing the environmental Kuznets curve hypothesis in Iceland: 1960−2010. *Revista de Economia del Rosario, 19*(1), 5−28.

Zambrano-Monserrate, M. A., Valverde-Bajana, I., Aguilar-Bohorquez, J., & Mendoza-Jimenez, M. (2016b). Relationship between economic growth and environmental degradation: Is there an environmental evidence of Kuznets curve for Brazil? *International Journal of Energy Economics and Policy, 6*(2), 208−216.

CHAPTER 5

The Process of Sustainability: From Past to Present

RECEP ULUCAK, PHD • ALI GÖKHAN YÜCEL, PHD • EMRAH KOÇAK, PHD

INTRODUCTION

Global warming has been a major issue of international agenda because of its crucial effects that have started to concretely appear around the world (Wang, Jiang, & Lang, 2017). Since the 1950s, observed changes such as decreases in amount of snow and ice, rises in sea levels and average temperature have been unprecedented, and these are the most powerful evidence for taking action (IPCC, 2014b). As underlined in the Paris Agreement, the rise in the global average temperature has been substantial, and it should be limited below 2°C relative to the preindustrial era.

Rises in global average temperature mostly stem from anthropogenic greenhouse gases (GHGs), which are caused by the combustion of fossil fuels (Jian-Bin, Shao-Wu, Yong, Zong-Ci, & Xin-Yu, 2012) to produce goods and services. On the other hand, solid and/or liquid wastes are produced while natural resources are depleted in the production process of goods and services. So, one might claim that the fundamental reasons of ecologic degradation that causes global warming, ozone layer depletion, and biodiversity loss are human activities such as production, consumption, and urbanization (Ulucak, 2018) that have dramatically increased since the 1950s. Observed changes in the nature have been realized in parallel with these activities, and they have raised concerns for sustainability worldwide. Then, sustainability issue has started to be widely discussed at both national and international levels. In this manner, sustainable development concept, which was firstly mentioned at the United Nations Conference in Stockholm (1972), has been often emphasized in summits organized by the United Nations to minimize environmental threats while countries sustain their economic growths within ecologic constraints.

Sustainable development concept is, basically, to meet today's needs without depriving ability for meeting needs of future. The concept has been built on three pillars: environmental, economic, and social aspects (Giddings, Hopwood, & O'Brien, 2002). Environmental aspect focuses on the maintenance of ecologic functions properly, whereas economic aspect aims at producing goods and services on a continuing basis. Social issues such as gender equity, health, education, democracy, judicial independence, accountability, and income inequality are the problem areas of the third pillar to achieve sustainable development (Harris, 2000). On the other hand, the concept has been discussed in weak and strong forms (Dietz & Neumayer, 2007; Garmendia, Prellezo, Murillas, Escapa, & Gallastegui, 2010; Neumayer, 2004, 2010; Victor, 2005). Weak sustainability is based on substitutability principle between natural capital (resources) and man-made capital, meaning that decreases in natural resources can be compensated by rises in man-made capital. By this way, sustainability can be achieved so long as total capital (natural capital + man-made capital) does not diminish over time (Neumayer, 2010). Strong sustainability is, however, against the substitutability between natural capital and man-made capital due to the fact that environmental threats caused by natural resources depletion and waste production might not be reversed (Martins, 2016; Oliveira Neto, Pinto, Amorim, Giannetti, & Almeida, 2018; Pearce & Barbier, 2000).

The emerging developments in environmental conditions have led economic thought to deep changes that fundamentally center on maintaining the healthy functioning of the ecosystem. Thus, new ecologic perspective on economics called the "ecologic economics" has started to become widespread by introducing alternative growth models, IS-LM analyses, and various macroeconomic implications (Costanza, Daly, & Bartholomew, 1991; Daly, 1991; Fontana & Sawyer, 2016; Hardt & O'Neill, 2017; Jackson & Victor, 2015; Rezai & Stagl, 2016; Rezai, Taylor, & Mechler, 2013;

Environmental Kuznets Curve (EKC). https//doi org/ 10. 1010/B978 0 12 816797 7 00005-9

Røpke, 2013; Stagl, 2014). On the other hand, growth strategies based on traditional economic thought have started to pay attention to environmental issues by including pollution, wastes, natural resources, renewables, and nonrenewables into their growth models (Acemoglu, Aghion, Bursztyn, & Hemous, 2012; Acemoglu, Akcigit, Hanley, & Kerr, 2016; Aghion, Dechezleprêtre, Hémous, Martin, & Van Reenen, 2016; Aghion & Howit, 1998; Barbier, 1999; Brock & Taylor, 2010; Reyes, 2011; Selden & Song, 1995; Stokey, 1998). It is important here to state a notice from Acemoglu et al. (2012) that traditional approach based on "laissez-faire equilibrium leads to an environmental disaster unless environmental quality cannot be sustained within a critical limit."

Because of increasing negative impacts of global warming, environmental sustainability has gained much importance and priority (Bilgili & Ulucak, 2018). However, economic concerns might preclude countries from putting strict environmental regulations into practice because they might restrict economic growth to a large extent. On the other hand, important arguments exist to support that increases in income might promote environmental protection. One argument indicates that the income elasticity of environment is greater than one. This elasticity output depicts that environmental goods are luxury, meaning that as income rises by 1% and the quantity demanded for environmental goods will increase by more than 1% (Baldwin, 1994). Furthermore, when one looks this relationship of growth-pollution from a different standpoint, he/she may claim that the poverty is a major cause and effect of global environmental problems as it is stressed in the Brundtland Report (1987) because citizens of poor countries excessively use environmental goods and this overuse exceeds regenerative capacity of environment (Yusuf, 2004). In addition, environmental degradation is evaluated as a major factor causing indigenous poverty (Duraiappah, 1996). In the light of these viewpoints, "Probably the only way to attain a decent environment in most countries is to become rich" (Beckerman, 1992).

There are other possible theoretic explanations to reveal why a rise in income lessens environmental degradation. They are *composition* and *technique* effects that make the relationship between income and pollution be inverted U-shape as hypothesized by Grossman and Krueger (1991, 1995). This relationship called the environmental Kuznets curve (EKC) has been a widely investigated research issue and verified by many studies (i.e., Acar & Aşıcı, 2016; Al-Mulali, Ozturk, & Solarin, 2016; Charfeddine & Mrabet, 2017; Ozcan, Apergis, & Shahbaz, 2018; Ozturk, Al-Mulali, & Saboori, 2016; Shahbaz, Solarin, Hammoudeh, & Shahzad, 2017; Shahbaz, Solarin, & Ozturk, 2016; Ulucak & Bilgili, 2018). Similarly, but following different model specifications, environmental convergence hypothesis also implies gradually decreasing path in terms of environmental degradation indicators as initial levels of relevant variables are associated with slower growth in convergence concept (Stern, 2015). Therefore, convergence approach for environmental degradation is proposed as an alternative to the EKC (Brock & Taylor, 2003, 2010; Stern, 2015). Convergence in various environmental degradation indicators has been investigated and important policy implications have been revealed by many studies (i.e., Acar & Lindmark, 2017; Burnett, 2016; Kounetas, 2018; Long, Sun, Cheng, & Zhang, 2017; Pennino, Bellido, Conesa, Coll, & Tortosa-Ausina, 2017; Ulucak & Apergis, 2018; Wesley Burnett & Madariaga, 2017).

Energy has become one the most important global issues as it plays a vital role in all three dimensions of sustainable development. For instance, (1) energy is an essential input for production that is the main driving element of economic growth regarding economic sustainability, (2) environmental problems are the focuses of environmental sustainability stem from production and consumption process of energy to a large extent, and (3) it can be considered as a basic human need that is required to meet to achieve social sustainability as well (Munasinghe, 2002; Najam & Cleveland, 2003). Therefore, energy has been a focal point of sustainable development summits and goals. To this end, policy implications have been discussed based on empirical findings in the relevant literature and energy policies have priority within these discussions. Implications and discussions in the literature reveal that renewable energy seems to be the best alternative for sustainable development (i.e., Al-mulali, Adebola Solarin, Sheau-Ting, & Ozturk, 2017; Bilgili, Koçak, Bulut, & Kuloğlu, 2017, Bilgili, Koçak, Bulut, & Kuşkaya, 2017; Dong et al., 2017; Ito, 2017; Kahia, Aïssa, & Lanouar, 2017; Liu, Zhang, & Bae, 2017) as was underlined by all international energy and environmental summits. Because renewable energy can be cost-efficient as well as eco-efficient for many countries in the long run due to stress of energy dependence and/or security problems (Hache, 2018; Kanchana, McLellan, & Unesaki, 2016; Proskuryakova, 2018; Wang, Wang, Wei, & Li, 2018).

This study aims at reviewing implications of energy-environmental summits and relevant literature in the context of sustainable development. In this regard, important developments in the diffusion of sustainable development approach at the global level are centered on by examining declarations of important summits and implications of the literature. The rest of the study is organized as follows: International summits and declarations on environment are presented in Section 2. Implications and suggestions from the literature are discussed in Section 3. The study is concluded with general observations on alternative solutions and suggestions.

FROM STOCKHOLM TO PARIS: A BRIEF HISTORY OF THE ENVIRONMENTAL AWARENESS

Global warming and environmental pollution are among the biggest threats the world is facing today. Environmental issues were not major concerns in the global agenda until the 1970s. Starting from the early 1970s, the global community has come a long way in environmental awareness. To fully understand the debate, one must look at the historic developments on environmental summits.

UN Conference on Human Development, Stockholm 1972

United Nations Conference on the Human Environment (UNCHE, also known as the Stockholm Conference) was first proposed to the United Nations by Sweden and was approved by the General Assembly in 1968. The conference was held from 5–16 June 1972 in Stockholm and was attended by 1200 official delegates from 113 countries. Even though there are earlier roots of environmental consciousness at political and academic level[1], for many, global awareness of environmental issues started with the Stockholm Conference. The conference attracted worldwide attention due to being "first" in many respects (Najam & Cleveland, 2003): (1) It was the first meeting that brought the nations together to address the environmental future of the world; (2) It became the first major conference of the United Nations that focused on only one issue; (3) It was the first global meeting that saw a large presence and influence of nonstate actors, including nongovernmental organizations (NGOs)

and scientists; and (4) It became the first meeting to search for global agreement on environment-related issues.

On the first day of the gathering, the World Environment Day was designated by the UN General Assembly to mark the opening of the conference, which is commemorated each year on 5 June. The gathering agreed on a declaration including 26 principles regarding the environment and development, an action plan with 109 recommendations and a resolution. The declaration jointly formulated by developed and developing countries is important in terms of being the first international agreement to recognize that both developing and developed countries contribute to environmental problems. The declaration noted that whereas the environmental problems in the developed countries are caused by industrialization and technologic development, in the developing countries, it is caused by underdevelopment.

Stockholm Conference might be regarded as a milestone in the international policy by formulating the concern of industrialized countries about environmental quality and the concern of developing countries about human development to a single composite framework, which would later be known as "sustainable development." The conference also led to the establishment of the United Nations Environment Program (UNEP), which serves as an authoritative advocate for the global environment.

Brundtland Report, 1987

Following the 10 years after the Stockholm conference, most of the global environmental issues were not adequately addressed. In contrast, the challenges such as pollution, acid rain, deforestation, and desertification even grew. In 1983, the United Nations established the World Commission on Environment and Development (WCED) to devise global strategies for protecting the environment and then appointed the Prime Minister of Norway, Gro Harlem Brundtland, as the chairperson of the commission. The Commission published its report entitled *Our Common Future*, later known as Brundtland report, in 1987. The report put forward the concept of "sustainable development" for the first time and described how it can be achieved.

In the report, sustainable development was defined as follows: *"Sustainable development is development that meets the needs of the present without compromising the ability of future generations to meet their own needs"* (WCED, 1987, p. 43), which later became the most quoted definition of the concept to date (Krueger & Agyeman, 2005). Brundtland report also highlighted

---

[1]Interested reader may refer to Worster (1994) for a comprehensive review of the studies on environment.

that economic development cannot be addressed separately from environmental protection.

Establishment of IPCC, 1988

As the topic of environment occupied the international agenda and the environmental awareness increased, some important steps were started to be taken. One of these steps was the formation of the Intergovernmental Panel on Climate Change (IPCC) by the UNEP and the World Meteorological Organization (WMO) in 1988 to present scientific information on climate change, its potential consequences and options for adaptation and mitigation. The IPCC published detailed assessments in 1990, 1995, 2001, 2007, and 2014. The fifth assessment report, AR5, finds, beyond reasonable doubt, that the Earth's climate is warming (IPCC, 2013).

UN Conference on Environment and Development, Rio de Janeiro 1992

"Our Common Future" report laid the foundation for the Rio Summit. Hosted by Rio de Janeiro from 3−14 June 1992 to coincide with the World Environment Day, June 5th, the United Nations Conference on Environment and Development (UNCED) became the biggest environmental conference ever organized with representatives from 178 nations and more than 30,000 participants including 108 heads of state and 2400 representatives of NGOs. The summit aimed to explore measures to reconcile economic activities with environmental protection to provide sustainable future for all individuals.

The outcomes of the UNCED summit, also known as the Earth Summit, were Agenda 21—an environmental action plan; the Rio Declaration on Environment and Development—a series of principles defining the rights and responsibilities of States; Principles of Forest Management—a set of principles to underlie the sustainable management of forests worldwide; the United Nations Framework Convention on Climate Change (UNFCCC)—an agreement on the climate change convention which later led to the Kyoto Protocol and the United Nations Convention on Biological Diversity (UNCBD).

Among the outcomes of Rio, Agenda 21[2], which is a nonbinding action plan of the United Nations with respect to sustainable development, was the major one. Officially stated, Agenda 21 is a *comprehensive plan of action to be taken globally, nationally and locally by organizations of the United Nations System, Governments, and Major Groups in every area in which human impacts on the environment"* (UNCED, 1992). The Earth Summit

also led to the establishment of the UN Commission on Sustainable Development (CSD) to assure follow-up of UNCED. Another significant impact of the Earth Summit was that, as the United Nations puts it: "it influenced all subsequent UN conferences, which have examined the relationship between human rights, population, social development, and the need for environmentally sustainable development".[3]

Kyoto Protocol, 1997

The adoption of UNFCCC was a significant step forward in challenging global warming. Yet as GHG emissions continued to increase all over the world, it became quite clear that only a strong and binding commitment by industrialized countries to decrease emissions could influence all the related parties and groups to participate in actions against climate change. Kyoto Protocol, named for the Japanese city where it was agreed on December 11, 1997—in full, the Protocol to the United Nations Framework Convention on Climate is one of the well-known and far-reaching plans undertaken by the UNFCCC. The protocol was opened for signature in March 1998 but came into force on February 16, 2005.[4]

The protocol was a momentous step forward in efforts to fight climate change by targeting to reduce the emission of gases that cause global warming. The protocol divides parties into three main groups as Annex I, Annex II, and non-Annex I countries according to different commitments.[5] During the first commitment period of 2008−2012, Annex I parties decided to reduce their overall emissions of GHG[6] by at least

---

[2]The "21" in Agenda 21 refers to the 21st century.

[3]http://www.un.org/geninfo/bp/enviro.html.
[4]The implementation of Kyoto Protocol has been delayed due to Article 25 which established a quite high threshold and employed 55/55 target. It not only requires ratification by 55 countries to the Convention, but also sufficient Annex I countries must ratify so as to represent not less than 55% of the total carbon dioxide emissions of Annex I countries in 1990. This requirement was only reached when Russia signed up for the Protocol in 2004, which was responsible for 17.4% of emissions in 1990.
[5]Annex I Parties are made up of industrialized countries that were members of the OECD (Organisation for Economic Cooperation and Development) in 1992 and economies in transition (the EIT Parties), including the Russian Federation, the Baltic States, and several Central and Eastern European States. Annex II Parties consist of the OECD members of Annex I, but not the EIT Parties. Non-Annex I Parties are mostly developing countries.
[6]By greenhouse gases, the Protocol includes six gases: "carbon dioxide, methane, nitrous oxide, hydrofluorocarbons, perfluorocarbons, and sulfur hexafluoride".

5% below 1990. Specific targets varied among the countries. Annex II parties are required to provide financial support to developing countries to help them reduce their GHG emissions and manage the impacts of climate change. Non-Annex I parties, mostly low-income countries, do not have legally binding emissions reductions targets to reduce or limit their emissions during the first commitment period.

Kyoto Protocol has introduced three flexibility mechanisms to help countries satisfy their emission requirements and to encourage businesses and low-income countries to contribute to emission reduction efforts.

Joint implementation (JI): As set out in Article 6 of the Kyoto Protocol, "a country with an emission reduction or limitation commitment under the Kyoto Protocol might take part in an emission reduction (or removal) project in any other country with a commitment under the Protocol, and it can credit the resulting emission units toward reaching its Kyoto target."

Clean development mechanism (CDM): Defined in Article 12 of the Kyoto Protocol, "the mechanism enables developing countries to carry out CDM projects to earn certified emission reduction credits. These credits could be traded and used by developed countries to meet their targets under the Kyoto Protocol."

International Emissions trading (IET): IET is the most well-known mechanism that allows trade of emission units between industrialized countries. According to the Article 17 of the Kyoto Protocol, "if a country achieves better reduction than it is required, it could sell surplus emission units to other countries. Likewise, if a country does not meet its Kyoto target, it could acquire these units and credit them to its own reduction."[7]

A second commitment period, an 8-year extension of the Kyoto Protocol from January 1, 2013, to December 31, 2020, was decided in the UN Doha Climate Change Conference in 2012, and 37 countries[8] had binding targets.

---

[7]This means that there will be a "carbon market" which would be tracked and traded like other commodities since carbon dioxide is the principal GHG.

[8]These countries are Australia, 28-member states of the European Union, Belarus, Iceland, Kazakhstan, Liechtenstein, Norway, Switzerland and Ukraine. Of these listed countries, Belarus, Kazakhstan and Ukraine stated that they might withdraw or not try to reach second-round targets. Russia, Japan and New Zealand participated in the first commitment of Kyoto but withdrew from the second round. Other industrialized countries that do not have second-round commitments include United States which did not ratify Kyoto Protocol and Canada which became the first country to pull out of Kyoto protocol in 2011 citing that the Protocol does not cover the two biggest carbon emitters, US and China.

World Summit on Sustainable Development, Johannesburg 2002

Damages in the earth's ecosystem at an unsustainable rate forced the world leaders to hold a summit to follow new initiatives on the implementation of sustainable development. Ten years after the first "Earth Summit" in Rio de Janeiro, the World Summit on Sustainable Development (WSSD), also known as the Earth Summit II or Rio + 10, was held in Johannesburg between August 26 and September 4, 2002. The conference was a follow-up on the implementation of the Rio 1992 conference, and the focus was sustainable development by the United Nations. The major outcome of Rio+10 was Johannesburg Declaration, which was built on previous declarations adopted at the UN conferences in Stockholm 1972 and Rio 1992. The Declaration consists of the aims and timetables for reaching the goals of Agenda 21.

United Nations Conference on Sustainable Development, Rio de Janeiro 2012

The United Nations Conference on Sustainable Development (UNCSD), also known as the Rio 2012, the Rio+20, or the Earth Summit 2012, was the third international conference on sustainable development targeting to reconcile the economic and environmental objectives of the global community. Held in Rio de Janeiro from 13 to 22 of June 2012, the conference was the follow-up of 1992 UNCED and 2002 WSSD. The conference did not produce any breakthrough agreements or commitments, yet it provided a platform to discuss sustainable development. The most significant outcome of the conference was the nonbinding document entitled "The Future We Want," which includes clear and practical ways to achieve the sustainable development. The conference also produced important guidelines on green economy policies.

United Nations Climate Change Conference, Paris 2015

As the second commitment of the Kyoto Protocol on GHG emissions will end in 2020, there is a need to discuss possible new agreements on climate change. Scientists' warning that if GHGs continue to increase, we will pass the threshold beyond which global warming becomes catastrophic and irreversible (Adger & Barnett, 2009; Allen et al., 2009; Charbit, Paillard, & Ramstein, 2008; Friedlingstein et al., 2014; Huntingford et al., 2012; Le Kama, Pommeret, & Prieur, 2014; Lowe et al., 2009; Matthews & Caldeira, 2008; Oberheitmann, 2013; Solomon et al., 2010) also made it clear for an international agreement to combat climate change. According to IPCC's fifth assessment report[9], limiting warming less than 2°C would require with a 66.7% probability of total carbon emissions since 1870 to remain below about 2900 gigatons. About two-thirds of this amount had already been emitted by 2011. To keep warming under 2°C, several hundred billion dollars each year should be invested in low carbon electricity and energy efficiency (IPCC, 2014a, p. 63).

2015 United Nations Climate Change Conference[10] was held from November 30 to December 12, 2015 in Paris. The conference negotiated Paris Agreement, an international treaty that aims to reduce the emission of GHGs. The Agreement was opened for signature on [11]April 22, 2016 and set a record for the first day signatures of support to an international agreement with 175 parties (174 countries and the EU) signing. The Agreement achieved the threshold[12] for entry into force on October 5, 2016 and came into force on November 5, 2016.

In Paris, countries from all over the world submitted their pledges to the United Nations, setting out how far they were willing to reduce their GHG emissions. Looking at the pledges of the biggest emitters, China ranks number one in the world with a share of 30% of global carbon emissions, committed to reach peak carbon emissions no later than 2030. The United States, second-largest emitter with a share of 15% of global carbon emissions, pledged to reduce its emissions by 26% −28% of 2005 levels by 2025. The European Union committed to at least 40% reduction in GHGs emissions by 2030 compared with 1990 levels. India vowed to reduce the emissions intensify relative to its GDP by 33%−35% from its 2005 levels by 2030.

The Agreement's main goal was to limit global temperature "well below 2° above pre-industrial levels and pursuing efforts to keep the temperature increase at 1.5°C above pre-industrial levels" since this would

notably decrease the risks and the effects of climate change.

Paris Agreement, which will replace the Kyoto Protocol from 2020, marked a breakpoint in the combat against climate change. The agreement has been the most comprehensive climate agreement to date. As of October 2018, the European Union and 194 countries signed the Agreement including top emitters; the United States, China, and India. Unlike its predecessor treaty, the Kyoto Protocol, the Paris deal represents a specific commitment to limit the increase in global temperatures "well below 2°C compared to the preindustrial times" and strive to stop at 1.5°C. The Agreement also sets a goal of delivering 100 billion USD each year in climate finance for developing countries by 2020, which would help poor countries to cope with the rising impacts of climate change.

On the negative aspects of the Paris Climate Agreement, two criticisms stand out. Firstly, major countries do not remain on track to meet their pledges to limit the gas emissions. According to a study published recently in *Nature* by David G. Victor et al. (2017), all major industrialized nations are failing short of their emissions reduction targets under the Paris Agreement. In addition, Rogelj and colleagues (Rogelj, et al., 2016) argue that even if these countries met their pledged emission reduction targets, it would not be adequate to keep global temperature rise well below 2°C. Secondly, the fact that the Agreement is nonbinding and that there are no penalties for noncompliance make it relatively hard to be successful. Because there is no penalty for leaving the deal, countries may quit the agreement any time. On June 1, 2017, President Trump stated that the United States, earth's second-largest polluter, would quit Paris deal by citing that the Agreement will hurt the US economy and put it at a permanent disadvantage, which became a heavy blow to global efforts against climate change.

Paris Agreement was the result of negotiations in the 2015 United Nations Climate Change Conference (COP 21/CMP 11). Looking at the subsequent conferences after Paris, the United Nations held two more climate change conferences in 2016 and 2017. The 22nd UN Climate Change Conference (COP 22) took place in Marrakech, November 7−18, 2016. The countries reaffirmed their commitment to the full implementation of the Paris Agreement. COP23, the second "conference of the parties" since the Paris Agreement, was held on 13−14 November in Bonn. It had been the first set of negotiations since the withdrawal of the United States from the Paris deal. Parties adopted 31 decisions on adaptation fund, gender, indigenous

---

[9]In Paris 2015, the IPCC was invited to publish a new special report (Special Report on 1.5 Degrees, SR15) in 2018 on the effects of global warming of 1.5°C above pre-industrial levels. The report was expected to be finalized in mid-October 2018.
[10]The conference is also known as COP21 due to being the 21st session of the Conference of the Parties (COP) to the 1992 UNFCC or CMP11 due to being the 11th session of the Meeting of the Parties (CMP) to the 1997 Kyoto Protocol.
[11]April 22 also marks the celebration of Earth Day.
[12]Similar to the threshold in Kyoto Protocol, the threshold is the ratification by not less than 55 countries accounting for not less than 55% of GHG emissions. It should be noted that while it is easy to reach to a minimum of 55 countries, it is not that easy to reach 55% of global emissions without the cooperation of the biggest emitters, China and the US.

people, and local communities. At the conference, Syria announced that it would sign the Paris Agreement, leaving the United States as the only country in the world to disavow the deal. Another important outcome at the COP was the launch of Powering Past Coal Alliance by 27 countries and regions to accelerate clean growth through the rapid phase-out of traditional coal power.

DISCUSSIONS ON ALTERNATIVE RECOMMENDATIONS

The literature on the factors affecting environmental quality is of great importance against combating global warming and climate change. Researchers have been examining the relationship between economic development, energy, and environment both in a theoretic and empirical framework and have presented many models, variables, and analytic methods. Since its proposal, EKC has been a dominant approach to modeling economic growth and environmental degradation (Stern, 2017). The EKC model postulates a reverse U-shaped relationship between economic growth and environmental degradation. Although the main idea of the model is that economic growth may cause pressure on the environment in the short term, the ongoing economic growth process will reduce environmental pollution in the long term. An important advantage of the EKC model is that it not only incorporates economic growth or income but also enables the estimation of the effect of other variables related to environment. In this way, researchers can investigate the impacts of many economic or noneconomic variables on the environment.

Reviewing the literature findings for the EKC concept, the following conclusions are reached: (1) Many studies support the EKC hypothesis for the relationship between economic growth and environment (Akbostancı, Türüt-Aşık, & Tunç, 2009; Cetin, Ecevit, & Yucel, 2018a; Farhani, Mrizak, Chaibi, & Rault, 2014; Fodha & Zaghdoud, 2010; Olale, Ochuodho, Lantz, & El Armali, 2018; Song, Zheng, & Tong, 2008).

The main recommendation of these studies is that policies that support economic growth will have a positive effect on the environment in the end. (2) Energy consumption is frequently used as a control variable to the EKC model in the literature. Some of these studies take total or fossil energy consumption into account (Cetin, Ecevit, & Yucel, 2018b; Dong et al., 2017; Lotfalipour, Falahi, & Ashena, 2010; Omri, 2013; Saboori & Sulaiman, 2013; Soytas & Sari, 2009) whereas others consider renewable energy consumption (Ben Jebli, Ben Youssef, & Ozturk, 2015; Bilgili, Koçak, & Bulut, 2016; Chen & Lei, 2018). Research findings

provide strong evidence that the use of fossil energy increases pollutant emissions, whereas the use of renewable energy reduces polluted gas emissions. (3) It has been observed that urbanization has been included as an important variable in the EKC models that recently explained the relationship between energy, economy, and environment. The literature findings show that the urbanization process drives such activities as industrialization, population growth, transportation sector, and trade in urban areas. Therefore, urbanization is recognized as an important factor that increases both energy density and pollutant gas emissions (Cetin et al., 2018a; He, Xu, Shen, Long, & Chen, 2017; Kasman & Duman, 2015; Ozturk & Al-Mulali, 2015; Rafiq, Salim, & Nielsen, 2016; Xu, Dong, & Yang, 2018). (4) Another strand of the literature focuses on economic variables such as financial development, trade, foreign direct investments (Cetin et al., 2018b; Farhani & Ozturk, 2015; Javid & Sharif, 2016; Koçak & Şarkgüneşi, 2018; Omri, Daly, Rault, & Chaibi, 2015), and noneconomic variables such as institutions, democracy and freedom in reducing environmental pollution (Adams & Klobodu, 2017; Joshi & Beck, 2018; Kashwan, 2017; Sulemana, James, & Rikoon, 2017). (5) Finally, there is substantial evidence that human capital and education are an important policy tool to fight against environmental threats in EKC models (Balaguer & Cantavella, 2018; Ulucak & Bilgili, 2018; F. Wang et al., 2018). Alternative recommendations to mitigate possible damages of environmental threats within the EKC concept bring energy policies to forefront. Particularly, renewable energy and energy efficiency gained much importance to deal with the global warming and the climate change properly.

Renewable Energy

Energy is the main production factor for economic growth and prosperity, and all kinds of production processes are necessarily related to energy. Today, 87% of global energy need is provided by fossil energy sources. On the other hand, the current energy system based on fossil energy sources is the major cause of global warming and climate change. Researchers have clearly demonstrated that nearly two-thirds of global GHG emissions are caused by fossil resources such as coal, oil, and natural gas. Today, it is clearly seen that the current energy system is not sustainable and needs to be changed.

Aforementioned focal points of environmental summits can be summarized as follows: (1) globally, CO_2 emissions should be reduced rapidly to combat global warming; (2) countries do not compromise economic growth and development targets; (3) the transition to the renewable/clean energy systems is an absolute necessity; and (4) energy efficiency plays a key role in the success of energy and environmental policies. Moreover, investments in energy efficiency, innovation, and technologic developments create a synergy for the renewable energy system (Figueres et al., 2017; Gusmão Caiado, Leal Filho, Quelhas, Luiz de Mattos Nascimento, & Ávila, 2018). At this point, renewable energy sources come into prominence as an important solution. In the examination of proposed solutions by environmental summits over the last 40 years, several questions have come to mind to be scrutinized: Is there a radical change in the energy structure at the global level? Can renewable energy sources compete with fossil sources? Are renewable energy sources economically feasible for developing countries? Addressing these questions is critical to long-term policies that will be formulated in a coherent and consistent manner.

First, there has been no radical structural change in the global energy supply system. According to Fig. 5.1, the share of fossil energy resources in the total supply was 87% and 86% for 2000 and 2016, respectively. Accordingly, the share of renewable energy in total energy supply has remained almost unchanged. On the other hand, Table 5.1 indicates that renewable energy production and fossil energy production have increased by 47% and 35%, respectively, in the last 15 years. It might be claimed that there is a notable increase in renewable energy production; however, it is easily realized that this increase did not make an appreciable change in the share of renewable. In addition, coal production, a huge source of pollutants, increased by 61% in the same period. Although developed countries have considerably reduced coal

use, it is still the most important source of energy in developing countries. For instance, today, about 45% of world coal production is carried out by China. India, Indonesia, and South Africa are the other important producer countries (IEA, 2017b). To sum up, there has been no considerable change in the world's dependence on fossil energy resources despite the increasing risk of global warming. Oil, natural gas, and coal are still the most dominant inputs of energy production as illustrated by Fig. 5.1.

Secondly, there has been considerable progress in competing renewable energy sources with fossil resources. In recent years, tremendous developments in renewable energy production capacity, investments, technology, distribution, and renewable energy market have drawn attention. For instance, worldwide electricity generation capacity from renewable energy sources increased by 50% for the period of 2000 to 2008. Germany, Spain, the United States, Japan, China, and India are the countries that have great share in this capacity, and they dominate the renewable energy production market (Arent, Wise, & Gelman, 2011). Moreover, while the investments in renewable energy sources were $163 billion in 2009, these investments increased to $242 billion in 2016. Whereas developed countries accounted for approximately 52% of renewable energy investments, developing countries had a share of 48% in investments. More importantly, the investment in renewable energy capacity over the last 5 years has been almost double the investment in fossil fuels (REN21, 2017). In recent years, it has been observed that governments have accelerated R&D investment and incentive policies in the field of renewable energy, and significant commercial breakthroughs have taken place. The private sector reacts to the incentives

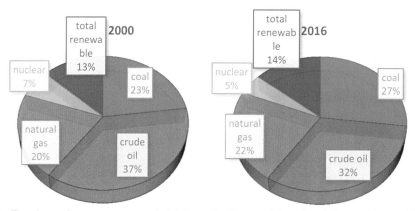

FIG. 5.1 The share of energy resources in total supply. (Source: International Energy Agency, Key World Energy Statistics.)

TABLE 5.1
World Total Primary Energy Supply by Source[a] (2000–16).

Source	2000	2005	2010	2016	% Change (2000–16)
Crude oil	3,746,765	4084,445	4,187,727	4482,631	20
Coal	2,316,125	2993,746	3,652,298	3730,866	61
Natural gas	2,072,073	2359,231	2,735,488	3034,954	46
Nuclear	675,467	721,726	718,829	679,649	0.6
Total fossil	8,810,430	10,159,148	11,294,342	11,928,100	35
Total renewable	1,307,487	1418,725	1,628,423	1924,139	47
Total primary energy supply (TPES)	10,117,917	11,577,873	12,922,765	13,852,239	37

[a] 1000 tonnes of oil equivalent (ktoe).
International Energy Agency, Key World Energy Statistics.

and policies of governments by accelerating their renewable R&D and commercial investments (Rausser, Stevens, & Torani, 2010). Many countries are increasing their subsidies for renewable energy. For example, since the 1970s, the United States has been implementing policies supporting biofuels. The US government has provided about $22 billion in support for biofuel production in 2010 (Bilgili, Koçak, Bulut, & Kuşkaya, 2017).

On the other hand, the decline in fossil fuel prices in recent years has had a negative impact on the competitiveness of renewable energy sources due to the fact that low oil prices make renewable energy sources less attractive from the point of their costs. In 2012, the average cost of energy resources per kWh in the United States (in cents) was 11.44 and 11.35 for coal and natural gas, respectively. Although the United States is one of the leading countries in renewable energy subsidies, investments, and technologies, the production costs for wind energy, photovoltaic (PV) solar, and thermal solar energy during the same period were 20.41, 13.00, and 24.31, respectively. Therefore, solar and wind energy in the United States do not have cost advantages compared with fossil energy. However, biomass, geothermal, and hydrogen energy are highly competitive with fossil energy (EIA, 2012). There has been a steady downward trend in renewable energy production costs in both the United States and the world. According to IRENA (2017) report, the global cost of renewable energy production has significantly decreased between 2010 and 2017. The decline in wind and solar electricity production costs were 50% and 20%, respectively. Developing countries have been experiencing a shortage of financing to expand their renewable capacities. Therefore, there is a lack of adequate investment and

technologic development performance in renewable materials. These adequacies negatively affect the competitiveness of renewable energy sources in developing countries. Cheaper coal prices also explain the increased usage of coal in developing countries.

It is clear that fossil fuels have a cost advantage over renewables in terms of market prices. However, economic theory suggests that environmental externalities should be added to the production costs of energy resources. The damage to the environment by fossil resources creates additional cost for societies, and this cost can be internalized by Pigouvian taxation that leads to adjustment of fossil energy prices. Thus, renewable energy sources can easily compete with fossils as long as economic agents are charged with social costs of their activities. In this way, it is more likely that renewable energy sources will become more effective sources in the long term (Timmons, Harris, & Roach, 2014).

Energy Efficiency
Energy efficiency is the reduction of the share of energy input in production or achievement of the same production level with less energy. Thus, energy consumption can be saved considerably through appropriate measures, and efficient usage of energy can be a key concept in the fight against global warming. The main measure of energy efficiency is energy intensity that is often used to compare countries' situations for fossil fuels. Energy intensity is calculated as the ratio of energy consumption to GDP, implying that how much energy is used to produce a unit of output. This indicator also provides a general idea of a country's industrial structure, technology level, and the energy use performance (Liu & Xie, 2013). Countries need to reduce energy intensity to achieve their long-term CO_2 emission

TABLE 5.2
Comparison of Energy Intensity[a] at Regional Levels.

	1990	2000	2005	2016
OECD	0.15	0.14	0.13	0.11
Non-OECD	0.47	0.38	0.37	0.29
Africa	0.42	0.43	0.39	0.35
Asia (exc. China)	0.44	0.40	0.37	0.29
China	0.99	0.51	0.50	0.31
Middle East	0.25	0.29	0.31	0.32
Non-OECD Americas	0.16	0.15	0.15	0.15
World	0.23	0.20	0.20	0.18

[a] TPES/GDP (Tonne of oil equivalent (toe)/1000 USD2010).
International Energy Agency, Key World Energy Statistics.

mitigation targets, and technologic advancements enable countries to get shrinkage in energy intensity. Therefore, developed countries that are also technologically advanced are superior to developing countries.

Table 5.2 shows a significant decrease in energy intensity over the last 30 years at the global level. For example, China, the world's biggest CO_2 emitter, has differentiated from other countries until 2000 in energy intensity. However, in recent years, China has considerably, if not sufficiently, reduced its energy intensity. The Middle East is the only region that has increased energy intensity in the last 30 years (see Table 5.2).

Determination of energy usage by sectors has a great matter to realize what policies should be followed to reduce energy intensity. According to the IEA (2017a) report, electricity and heating, transportation, industry, and residential sectors account for 42%, 24%, 19%, and 6% of total CO_2 emissions from fossil energy, respectively. One might easily conclude that areas with high energy intensity are the modern cities where industry, trade, and job opportunities are high and migration and transportation are developed (Bilgili, Koçak, Bulut, & Kuloğlu, 2017). It should be noted that the pressure on cities' energy intensity is balanced by free market conditions because the free market directs urban production to be more innovative and encourages companies to use more modern-environment-friendly technologies (Madlener & Sunak, 2011).

Today, developed countries provide some practices in urban areas that will reduce the energy intensity caused by the transportation sector. Developing countries lack such an application due to infrastructure deficiencies (Madlener & Sunak, 2011). Similarly,

encouraging the use of electric vehicles and biofuels to replace oil use in the transport sector will reduce fossil energy use (Bilgili, Koçak, Bulut, & Kuloğlu, 2017; Figueres et al., 2017).

High-rise buildings in the urban areas, infrastructure, and construction sectors are the other activities that increase energy intensity. Green building codes, such as the recently developed Leadership of Energy and Environmental Design (LEED) certificates, reduce the energy intensity in buildings (Sadorsky, 2013). Especially in developing countries, there is no standard in terms of energy efficiency in buildings due to rapid and unbalanced urbanization. Therefore, the construction of new energy-efficient buildings is of importance and might be an alternative solution to decrease energy usage and intensity in developing countries.

Non–Energy Policy Recommendations
The following policies have been discussed in the literature to balance CO_2 emissions among countries or companies and to mitigate environmental stresses:

- **Carbon market and carbon trading:** The carbon market refers to the market in which carbon credits, in other words carbon certificates, are obtained and sold within defined standards for the prevention or reduction of GHGs. This market punishes businesses that emit more than the limit, while rewarding those who emit less. Thus, the carbon market provides environmental efficiency in the use of resources. Although the value of the European Union Emissions Trading market was $11 billion in 2005, it reached $176 billion in 2011 (Narin, 2013). This increase in 6 years is promising for the future of the carbon market because it can encourage businesses and countries to use clean technology.

- **Land-use policy:** Land-use policies focus on reduction in forest damages and supporting afforestation because forests have the ability to capture and store carbon emissions (Bayrac & Dogan, 2016). Such policies should not be ignored to prevent environmental threats due to emissions stemming from deforestation and land-use changes have a considerable share (12%) in global GHG emissions.

- **Financing of climate change mitigation efforts:** Governments, private banks, and the World Bank might support green activities against the threats. For instance, these organizations provided about $81 billion in funding in 2016 to combat climate change. However, such organizations must export more "green bonds" to finance climate change projects (Figueres et al., 2017) and to expand green efforts further.

- **Education:** Education creates awareness that can help improve environmental quality (Balaguer & Cantavella, 2018) and contributes to increase in the number of talented individuals for the development of clean technologies. The participation of the whole society in the activities that aim at preserving environmental values is of prime importance in policy design and implementation. Therefore, education, training, and public awareness on climate change should be regarded as an important policy tool (Anderson, 2012).

CONCLUSION

Based on observed changes in amount of snow and ice, rises in sea levels as well as measured global average temperature, there has been a considerable awareness about environmental threats in global summits under the guidance of the UN. Not only these observed and measured changes but also natural resources and biodiversity that have crucial roles in ecosystems are under threat. For instance, the world ecologic footprint that measures people's demand for nature has exceeded biocapacity that represents ecologic assets since the 1970s. According to Global Footprint Network calculations, by overharvesting resources and emitting more CO_2 than regenerative capacity can sequester, people exhaust 1.7 more natural assets than they can be regenerated by nature, and this is an unsustainable benchmark for a healthy ecosystem (Wackernagel, Lin, Hanscom, Galli, & Iha, 2019). On the other hand, there is a broad consensus about substantial increase in biodiversity loss that may cause destructive effects on ecosystems (Skogen, Helland, & Kaltenborn, 2018). Therefore, it is an indisputable fact that the required actions must be put into practice to be able to survive in the world without delay.

The Stockholm Conference (1972), Brundtland Report (1987), Rio Conference (1992), Kyoto Protocol (1997), Rio+10 in Johannesburg (2002), Rio+20 in Rio (2012), Paris Climate Conference (2015), and the annual periodical assessment meetings of Conference of Parties (COP1—COP23) are important efforts and common platform for countries to evoke actions for sustainable development. However, a noteworthy progress has not been experienced so far because most emitted countries such as the United States, China, and India have not taken responsibility adequately in the first commitment period of Kyoto (2008—12). Moreover, the United States has not signed Kyoto. As a result, the target of 5% reduction in CO_2 emission levels of the 1990s for the first commitment period could not be achieved by many countries (see Kuriyama

& Abe, 2018), and CO_2 emissions continued to increase dramatically. Then, second commitment period has started for 2012—20 in Doha (COP18). It seems that the second period will end with failure under the same conditions (Almer & Winkler, 2017). Therefore, the requirement of actions that are more efficient and well attended has emerged since the motivation of countries cannot be strengthened to struggle with threats due to some deficiencies of Kyoto. In this regard, Paris Climate Agreement was signed by 196 countries, and each country submitted their own abatement targets (INDCs) to be able to keep the increase in global average temperature within the critical threshold.

Paris Climate Agreement was evaluated as a "big deal" and "greatest diplomatic success" in the world by international press[13] at first. However, it is scientifically projected that commitments are not enough to keep global average temperature below 2 °C from preindustrial levels (CI, 2018; Oberthür & Groen, 2018; Rogelj et al., 2016; The Lancet Planetary Health, 2018; F. Wang et al., 2018). Therefore, one might claim that Paris pledges are not sufficient to achieve the aim of staying under 2°C, even if all countries pull their weights properly. On the other hand, the US withdrawal from the agreement, as one of the most emitted countries, may lead to motivation loss in developing countries to fulfill their commitments. Fortunately, 17 states joined the Climate Alliance that was founded on June 1, 2017 on the withdrawal of the United States and committed to Paris Agreement. The Alliance states have concerns about climate change, and they have decided to prevent environmental threats by acting as a part of the agreement. In addition, more than 200 city mayors and a large number of companies declared to fight environmental threats as is routed by Paris Agreement in the United States (Zhang, Chao, Zheng, & Huang, 2017). At least, these individual efforts of states, companies, and NGOs in the United States might positively be perceived by other countries and make the agreement worth being pursued. Even so, more efforts should be made to extend available pledges of Paris Agreement and to implement reasonable sanctions by the United Nations and leading countries.

All summits focusing on environmental issues give special attention to energy because it is directly related to three pillars of sustainable development

---

[13]https://www.nytimes.com/2015/12/16/opinion/paris-climate-accord-is-a-big-big-deal.html?smprod=nytcore-ipad&smid=nytcore-ipad-share&_r=0 https://www.theguardian.com/environment/2015/dec/13/paris-climate-deal-cop-diplomacy-developing-united-nations.

(Munasinghe, 2002; Najam & Cleveland, 2003). Renewable energy has top priority as an alternative policy recommendation, and empirical researches strongly support that countries will have the cost advantage in the long run by transforming energy systems with renewable ones. Because once renewable energy infrastructures are sufficiently advanced and clean technology standards are diffused, there will be no need for governmental incentives or interventions any more in the long run as is underlined by Acemoglu et al. (2012). Indeed, countries may bear the cost for renewable energy diffusion that considerably decreased in 2010−17 (IRENA, 2017); they will have the chance of living in better conditions and will not suffer from energy supply security and dependence problems in the long run. Similarly, energy efficiency that is mostly based on clean technology advancement is another solution to mitigate emissions. In this regard, urbanization and transportation that have important share of energy consumption should not be neglected in emission mitigation, and they are promoted and/or imposed to reach minimum quality standard for clean development by following advanced samples in sustainable cities and transportations. Otherwise, *our alternative is a dark and dangerous future* as stated by the Secretary-General of the United Nations.

REFERENCES

Acar, S., & Aşıcı, A. A. (2016). Does income growth relocate ecological footprint? *Ecological Indicators, 61*, 707−714. https://doi.org/10.1016/j.ecolind.2015.10.022.

Acar, S., & Lindmark, M. (2017). Convergence of CO_2 emissions and economic growth in the OECD countries: Did the type of fuel matter? *Energy Sources, Part B: Economics, Planning and Policy*, 1−10. https://doi.org/10.1080/15567249.2016.1249807.

Acemoglu, D., Aghion, P., Bursztyn, L., & Hemous, D. (2012). The environment and directed technical change. *The American Economic Review, 102*(1), 131−166.

Acemoglu, D., Akcigit, U., Hanley, D., & Kerr, W. (2016). Transition to clean technology. *Journal of Political Economy, 124*(1), 52−104. https://doi.org/10.1086/684511.

Adams, S., & Klobodu, E. K. M. (2017). Urbanization, democracy, bureaucratic quality, and environmental degradation. *Journal of Policy Modeling, 39*(6), 1035−1051. https://doi.org/10.1016/J.JPOLMOD.2017.04.006.

Adger, W. N., & Barnett, J. (2009). Four reasons for concern about adaptation to climate change. *Environment & Planning A, 41*(12), 2800−2805. https://doi.org/10.1068/a42244.

Aghion, P., Dechezleprêtre, A., Hémous, D., Martin, R., & Van Reenen, J. (2016). Carbon taxes, path dependency, and directed technical change: Evidence from the auto industry. *Journal of Political Economy, 124*(1), 1−51. https://doi.org/10.1086/684581.

Aghion, P., & Howit, P. (1998). *Endogenous growth theory*. Massachusetts: MIT Press.

Akbostancı, E., Turut-Asik, S., & Tunc, G.İ. (2009). The relationship between income and environment in Turkey: Is there an environmental Kuznets curve? *Energy Policy, 37*(3), 861−867. https://doi.org/10.1016/J.ENPOL.2008.09.088.

Al-mulali, U., Adebola Solarin, S., Sheau-Ting, L., & Ozturk, I. (2017). Corrigendum to "Does moving towards renewable energy cause water and land inefficiency? An empirical investigation. *Energy Policy, 111*, 297 [Energy Policy 93 (2016) 303−314] https://doi.org/10.1016/J.ENPOL.2017.08.045.

Al-Mulali, U., Ozturk, I., & Solarin, S. A. (2016). Investigating the environmental Kuznets curve hypothesis in seven regions: The role of renewable energy. *Ecological Indicators, 67*, 267−282. https://doi.org/10.1016/j.ecolind.2016.02.059.

Allen, M. R., Frame, D. J., Huntingford, C., Jones, C. D., Lowe, J. A., Meinshausen, M., et al. (2009). Warming caused by cumulative carbon emissions towards the trillionth tonne. *Nature, 458*(7242), 1163−1166. https://doi.org/10.1038/nature08019.

Almer, C., & Winkler, R. (2017). Analyzing the effectiveness of international environmental policies: The case of the Kyoto Protocol. *Journal of Environmental Economics and Management, 82*, 125−151. https://doi.org/10.1016/J.JEEM.2016.11.003.

Anderson, A. (2012). Climate change education for mitigation and adaptation. *Journal of Education for Sustainable Development, 6*(2), 191−206. https://doi.org/10.1177/0973408212475199.

Arent, D. J., Wise, A., & Gelman, R. (2011). The status and prospects of renewable energy for combating global warming. *Energy Economics, 33*(4), 584−593. https://doi.org/10.1016/J.ENECO.2010.11.003.

Balaguer, J., & Cantavella, M. (2018). The role of education in the environmental Kuznets Curve. Evidence from Australian data. *Energy Economics, 70*, 289−296. https://doi.org/10.1016/J.ENECO.2018.01.021.

Baldwin, R. (1994). Does sustainability require growth? In I. Goldin & L. A. Winters (Eds.), *The economics of sustainable development* (pp. 51−78). Cambridge: Cambridge University Press. First.

Barbier, E. B. (1999). Endogenous growth and natural resource scarcity. *Environmental and Resource Economics, 14*(1), 51−74. https://doi.org/10.1023/A:1008389422019.

Bayrac, H. N., & Dogan, E. (2016). Türkiye'de İklim Değişikliğinin Tarım Sektörü Üzerine Etkileri. *Eskişehir Osmangazi Üniversitesi İktisadi ve İdari Bilimler Fakültesi Dergisi, 11*(1), 23−48.

Beckerman, W. (1992). Economic growth and the environment: Whose growth? whose environment? *World Development, 20*(4), 481−496. https://doi.org/10.1016/0305-750X(92)90038-W.

Ben Jebli, M., Ben Youssef, S., & Ozturk, I. (2015). The role of renewable energy consumption and trade: Environmental Kuznets curve analysis for Sub-Saharan Africa Countries. *African Development Review, 27*(3), 288−300. https://doi.org/10.1111/1467-8268.12147.

Bilgili, F., Koçak, E., & Bulut, Ü. (2016). The dynamic impact of renewable energy consumption on CO_2 emissions: A revisited environmental Kuznets Curve approach. *Renewable and Sustainable Energy Reviews, 54,* 838–845. https://doi.org/10.1016/J.RSER.2015.10.080.

Bilgili, F., Koçak, E., Bulut, Ü., & Kuloğlu, A. (2017). The impact of urbanization on energy intensity: Panel data evidence considering cross-sectional dependence and heterogeneity. *Energy, 133,* 242–256. https://doi.org/10.1016/J.ENERGY.2017.05.121.

Bilgili, F., Koçak, E., Bulut, Ü., & Kuşkaya, S. (2017). Can biomass energy be an efficient policy tool for sustainable development? *Renewable and Sustainable Energy Reviews, 71,* 830–845. https://doi.org/10.1016/j.rser.2016.12.109.

Bilgili, F., & Ulucak, R. (2018). The nexus between biomass – footprint and sustainable development. In *Reference Module in Materials Science and Materials Engineering.* Elsevier. https://doi.org/10.1016/B978-0-12-803581-8.10600-9.

Brock, W. A., & Taylor, M. S. (2003). *The kindergarten rule of sustainable growth* (NBER No. Working Paper 9597). Massachusetts.

Brock, W. A., & Taylor, M. S. (2010). The Green Solow model. *Journal of Economic Growth, 15*(2), 127–153. https://doi.org/10.1007/s10887-010-9051-0.

Burnett, J. W. (2016). Club convergence and clustering of U.S. energy-related CO_2 emissions. *Resource and Energy Economics, 46,* 62–84. https://doi.org/10.1016/j.reseneeco.2016.09.001.

Burnett, W. J., & Madariaga, J. (2017). The convergence of U.S. state-level energy intensity. *Energy Economics, 62,* 357–370. https://doi.org/10.1016/J.ENECO.2016.03.029.

Cetin, M., Ecevit, E., & Yucel, A. G. (2018a). Structural breaks, urbanization and CO_2 emissions: Evidence from Turkey. *Journal of Applied Economics and Business Research, 8*(2), 122–139. Retrieved from http://www.aebrjournal.org/uploads/6/6/2/2/6622240/joaebrjune2018_122_139.pdf.

Cetin, M., Ecevit, E., & Yucel, A. G. (2018b). The impact of economic growth, energy consumption, trade openness, and financial development on carbon emissions: empirical evidence from Turkey. *Environmental Science and Pollution Research.* https://doi.org/10.1007/s11356-018-3526-5.

Charbit, S., Paillard, D., & Ramstein, G. (2008). Amount of CO_2 emissions irreversibly leading to the total melting of Greenland. *Geophysical Research Letters, 35*(12). n/a-n/a https://doi.org/10.1029/2008GL033472.

Charfeddine, L., & Mrabet, Z. (2017). The impact of economic development and social-political factors on ecological footprint: A panel data analysis for 15 MENA countries. *Renewable and Sustainable Energy Reviews, 76,* 138–154. https://doi.org/10.1016/j.rser.2017.03.031.

Chen, W., & Lei, Y. (2018). The impacts of renewable energy and technological innovation on environment-energy-growth nexus: New evidence from a panel quantile regression. *Renewable Energy, 123,* 1–14. https://doi.org/10.1016/J.RENENE.2018.02.026.

CI. (2018). *Climate scoreboard, UN climate pledge analysis.*

Costanza, R., Daly, H. E., & Bartholomew, J. A. (1991). Goals, agenda, and policy recommendations for ecological economics. In R. Costanza (Ed.), *Ecological economics: The science and management of sustainability* (p. 525). New York: Columbia University Press.

Daly, H. E. (1991). Elements of environmental of macroeconomics. In R. Costanza (Ed.), *Ecological economics: The science and management of sustainability.* New York: Columbia University Press.

Dietz, S., & Neumayer, E. (2007). Weak and strong sustainability in the SEEA: Concepts and measurement. *Ecological Economics, 61*(4), 617–626. https://doi.org/10.1016/j.ecolecon.2006.09.007.

Dong, K., Sun, R., Hochman, G., Zeng, X., Li, H., & Jiang, H. (2017). Impact of natural gas consumption on CO_2 emissions: Panel data evidence from China's provinces. *Journal of Cleaner Production, 162,* 400–410. https://doi.org/10.1016/J.JCLEPRO.2017.06.100.

Duraiappah, A. (1996). *Poverty and environmental degradation: A literature review and analysis.* No. 8.

EIA. (2012). *Annual energy outlook 2012: with projections to 2035.* Retrieved from: https://www.eia.gov/outlooks/aeo/pdf/0383(2012).pdf.

Farhani, S., Mrizak, S., Chaibi, A., & Rault, C. (2014). The environmental Kuznets curve and sustainability: A panel data analysis. *Energy Policy, 71,* 189–198. https://doi.org/10.1016/J.ENPOL.2014.04.030.

Farhani, S., & Ozturk, I. (2015). Causal relationship between CO_2 emissions, real GDP, energy consumption, financial development, trade openness, and urbanization in Tunisia. *Environmental Science and Pollution Research, 22*(20), 15663–15676. https://doi.org/10.1007/s11356-015-4767-1.

Figueres, C., Schellnhuber, H. J., Whiteman, G., Rockström, J., Hobley, A., & Rahmstorf, S. (2017). Three years to safeguard our climate. *Nature, 546*(7660), 593–595. https://doi.org/10.1038/546593a.

Fodha, M., & Zaghdoud, O. (2010). Economic growth and pollutant emissions in Tunisia: An empirical analysis of the environmental Kuznets curve. *Energy Policy, 38*(2), 1150–1156. https://doi.org/10.1016/J.ENPOL.2009.11.002.

Fontana, G., & Sawyer, M. (2016). Towards post-Keynesian ecological macroeconomics. *Ecological Economics, 121,* 186–195. https://doi.org/10.1016/J.ECOLECON.2015.03.017.

Friedlingstein, P., Andrew, R. M., Rogelj, J., Peters, G. P., Canadell, J. G., Knutti, R., et al. (2014). Persistent growth of CO_2 emissions and implications for reaching climate targets. *Nature Geoscience, 7*(10), 709–715. https://doi.org/10.1038/ngeo2248.

Garmendia, E., Prellezo, R., Murillas, A., Escapa, M., & Gallastegui, M. (2010). Weak and strong sustainability assessment in fisheries. *Ecological Economics, 70*(1), 96–106. https://doi.org/10.1016/j.ecolecon.2010.08.001.

Giddings, B., Hopwood, B., & O'Brien, G. (2002). Environment, economy and society: fitting them together into sustainable development. *Sustainable Development, 10*(4), 187–196. https://doi.org/10.1002/sd.199.

Grossman, G. M., & Krueger, A. B. (1991). *Environmental impacts of a North American free trade agreement.* National Bureau of Economic Research Working Paper Series, No. 3914(3914) https://doi.org/10.3386/w3914.

Grossman, G. M., & Krueger, A. B. (1995). Economic growth and the environment. *The Quarterly Journal of Economics, 110*(2), 353–377. https://doi.org/10.2307/2118443.

Gusmão Caiado, R. G., Leal Filho, W., Quelhas, O. L. G., Luiz de Mattos Nascimento, D., & Ávila, L. V. (2018). A literature-based review on potentials and constraints in the implementation of the sustainable development goals. *Journal of Cleaner Production, 198,* 1276–1288. https://doi.org/10.1016/J.JCLEPRO.2018.07.102.

Hache, E. (2018). Do renewable energies improve energy security in the long run? *International Economics.* https://doi.org/10.1016/J.INTECO.2018.01.005.

Hardt, L., & O'Neill, D. W. (2017). Ecological macroeconomic models: Assessing current developments. *Ecological Economics, 134,* 198–211. https://doi.org/10.1016/J.ECOLECON.2016.12.027.

Harris, J. M. (2000). *Basic principles of sustainable development* (No. WP00-04). Massachusetts.

He, Z., Xu, S., Shen, W., Long, R., & Chen, H. (2017). Impact of urbanization on energy related CO_2 emission at different development levels: Regional difference in China based on panel estimation. *Journal of Cleaner Production, 140,* 1719–1730. https://doi.org/10.1016/J.JCLEPRO.2016.08.155.

Huntingford, C., Lowe, J. A., Gohar, L. K., Bowerman, N. H. A., Allen, M. R., Raper, S. C. B., et al. (2012). The link between a global 2 °C warming threshold and emissions in years 2020, 2050 and beyond. *Environmental Research Letters, 7*(1), 014039. https://doi.org/10.1088/1748-9326/7/1/014039.

International Energy Agency. (2017a). *CO_2 emissions from fuel combustion 2017 overview.* Retrieved from: https://webstore.iea.org/co2-emissions-from-fuel-combustion-overview-2017.

International Energy Agency. (2017b). *Coal information 2017: Overview.* Retrieved from: https://webstore.iea.org/coal-information-2017-overview.

IPCC. (2013). *Climate change 2013 headline statements from the summary for policymakers.*

IPCC. (2014a). Climate change 2014: Synthesis report. In *Contribution of working groups I, II, and III to the fifth assesment report of the intergovernmental panel on the climate change.*

IPCC. (2014b). Synthesis report. In *Contribution of working groups I, II and III to the fifth assessment report of the intergovernmental panel on climate change. Geneva, Switzerland.*

IRENA. (2017). *Renewable power generation costs in 2017.* Retrieved from: http://www.irena.org/publications/2018/Jan/Renewable-power-generation-costs-in-2017.

Ito, K. (2017). CO_2 emissions, renewable and non-renewable energy consumption, and economic growth: Evidence from panel data for developing countries. *International Economics, 151,* 1–6. https://doi.org/10.1016/J.INTECO.2017.02.001.

Jackson, T., & Victor, P. A. (2015). Does credit create a 'growth imperative'? A quasi-stationary economy with interest-bearing debt. *Ecological Economics, 120,* 32–48. https://doi.org/10.1016/J.ECOLECON.2015.09.009.

Javid, M., & Sharif, F. (2016). Environmental Kuznets curve and financial development in Pakistan. *Renewable and Sustainable Energy Reviews, 54,* 406–414. https://doi.org/10.1016/J.RSER.2015.10.019.

Jian-Bin, H., Shao-Wu, W., Yong, L., Zong-Ci, Z., & Xin-Yu, W. (2012). Debates on the causes of global warming. *Advances in Climate Change Research, 3*(1), 38–44. https://doi.org/10.3724/SP.J.1248.2012.00038.

Joshi, P., & Beck, K. (2018). Democracy and carbon dioxide emissions: Assessing the interactions of political and economic freedom and the environmental Kuznets curve. *Energy Research & Social Science, 39,* 46–54. https://doi.org/10.1016/J.ERSS.2017.10.020.

Kahia, M., Aïssa, M. S. B., & Lanouar, C. (2017). Renewable and non-renewable energy use – economic growth nexus: The case of MENA net oil importing countries. *Renewable and Sustainable Energy Reviews, 71,* 127–140. https://doi.org/10.1016/J.RSER.2017.01.010.

Kanchana, K., McLellan, B. C., & Unesaki, H. (2016). Energy dependence with an Asian twist? Examining international energy relations in Southeast Asia. *Energy Research & Social Science, 21,* 123–140. https://doi.org/10.1016/J.ERSS.2016.07.003.

Kashwan, P. (2017). Inequality, democracy, and the environment: A cross-national analysis. *Ecological Economics, 131,* 139–151. https://doi.org/10.1016/J.ECOLECON.2016.08.018.

Kasman, A., & Duman, Y. S. (2015). CO_2 emissions, economic growth, energy consumption, trade and urbanization in new EU member and candidate countries: A panel data analysis. *Economic Modelling, 44,* 97–103. https://doi.org/10.1016/J.ECONMOD.2014.10.022.

Kounetas, K. E. (2018). Energy consumption and CO_2 emissions convergence in European Union member countries. A tonneau des Danaides? *Energy Economics, 69,* 111–127. https://doi.org/10.1016/j.eneco.2017.11.015.

Koçak, E., & Şarkgüneşi, A. (2018). The impact of foreign direct investment on CO_2 emissions in Turkey: new evidence from cointegration and bootstrap causality analysis. *Environmental Science and Pollution Research, 25*(1), 790–804. https://doi.org/10.1007/s11356-017-0468-2.

Krueger, R., & Agyeman, J. (2005). Sustainability schizophrenia or "actually existing sustainabilities?" toward a broader understanding of the politics and promise of local sustainability in the US. *Geoforum, 36*(4), 410–417. https://doi.org/10.1016/J.GEOFORUM.2004.07.005.

Kuriyama, A., & Abe, N. (2018). Ex-post assessment of the Kyoto Protocol – quantification of CO_2 mitigation impact in both Annex B and non-Annex B countries-. *Applied Energy, 220,* 286–295. https://doi.org/10.1016/J.APENERGY.2018.03.025.

Le Kama, A. A., Pommeret, A., & Prieur, F. (2014). Optimal emission policy under the risk of irreversible pollution.

Journal of Public Economic Theory, 16(6), 959–980. https://doi.org/10.1111/jpet.12085.

Liu, Y., & Xie, Y. (2013). Asymmetric adjustment of the dynamic relationship between energy intensity and urbanization in China. *Energy Economics, 36*, 43–54. https://doi.org/10.1016/J.ENECO.2012.12.003.

Liu, X., Zhang, S., & Bae, J. (2017). The impact of renewable energy and agriculture on carbon dioxide emissions: Investigating the environmental Kuznets curve in four selected ASEAN countries. *Journal of Cleaner Production, 164*, 1239–1247. https://doi.org/10.1016/J.JCLEPRO.2017.07.086.

Long, X., Sun, M., Cheng, F., & Zhang, J. (2017). Convergence analysis of eco-efficiency of China's cement manufacturers through unit root test of panel data. *Energy, 134*, 709–717. https://doi.org/10.1016/J.ENERGY.2017.05.079.

Lotfalipour, M. R., Falahi, M. A., & Ashena, M. (2010). Economic growth, CO_2 emissions, and fossil fuels consumption in Iran. *Energy, 35*(12), 5115–5120. https://doi.org/10.1016/J.ENERGY.2010.08.004.

Lowe, J. A., Huntingford, C., Raper, S. C. B., Jones, C. D., Liddicoat, S. K., & Gohar, L. K. (2009). How difficult is it to recover from dangerous levels of global warming? *Environmental Research Letters, 4*(1), 014012 https://doi.org/10.1088/1748-9326/4/1/014012.

Madlener, R., & Sunak, Y. (2011). Impacts of urbanization on urban structures and energy demand: What can we learn for urban energy planning and urbanization management? *Sustainable Cities and Society, 1*(1), 45–53. https://doi.org/10.1016/J.SCS.2010.08.006.

Martins, N. O. (2016). Ecosystems, strong sustainability and the classical circular economy. *Ecological Economics, 129*, 32–39. https://doi.org/10.1016/j.ecolecon.2016.06.003.

Matthews, H. D., & Caldeira, K. (2008). Stabilizing climate requires near-zero emissions. *Geophysical Research Letters, 35*(4), L04705. https://doi.org/10.1029/2007GL032388.

Munasinghe, M. (2002). The sustainomics trans-disciplinary meta-framework for making development more sustainable: applications to energy issues. *International Journal of Sustainable Development, 5*(1/2), 125. https://doi.org/10.1504/IJSD.2002.002563.

Najam, A., & Cleveland, C. J. (2003). Energy and sustainable development at global environmental summits: An evolving agenda. *Environment, Development and Sustainability, 5*(1/2), 117–138. https://doi.org/10.1023/A:1025388420042.

Narin, M. (2013). Kyoto Protokolü Esneklik Mekanizması: Emisyon Ticareti. In *Flexible mechanisms of the Kyoto protocol: emissions trading*. Retrieved from: https://www.avekon.org/papers/770.pdf.

Neumayer, E. (2004). Indicators of sustainability. In T. H. Tietenberg, & H. Folmer (Eds.), *The international yearbook of environmental and resource economics 2004/2005* (pp. 139–189). Cheltenham: Edward Elgar.

Neumayer, E. (2010). *Weak versus strong sustainability : exploring the limits of two opposing paradigms* (3rd ed.). Cheltenham, UK and Northampton MA, USA: Edward Elgar.

Oberheitmann, A. (2013). Some remarks on the individual contribution to climate change. *American Journal of Climate Change, 02*(03), 198–202. https://doi.org/10.4236/ajcc.2013.23020.

Oberthür, S., & Groen, L. (2018). Explaining goal achievement in international negotiations: the EU and the Paris agreement on climate change. *Journal of European Public Policy, 25*(5), 708–727. https://doi.org/10.1080/13501763.2017.1291708.

Olale, E., Ochuodho, T. O., Lantz, V., & El Armali, J. (2018). The environmental Kuznets curve model for greenhouse gas emissions in Canada. *Journal of Cleaner Production, 184*, 859–868. https://doi.org/10.1016/J.JCLEPRO.2018.02.178.

Oliveira Neto, G. C. de, Pinto, L. F. R., Amorim, M. P. C., Giannetti, B. F., & Almeida, C. M. V. B. de (2018). A framework of actions for strong sustainability. *Journal of Cleaner Production, 196*, 1629–1643. https://doi.org/10.1016/j.jclepro.2018.06.067.

Omri, A. (2013). CO_2 emissions, energy consumption and economic growth nexus in MENA countries: Evidence from simultaneous equations models. *Energy Economics, 40*, 657–664. https://doi.org/10.1016/J.ENECO.2013.09.003.

Omri, A., Daly, S., Rault, C., & Chaibi, A. (2015). Financial development, environmental quality, trade and economic growth: What causes what in MENA countries. *Energy Economics, 48*, 242–252. https://doi.org/10.1016/J.ENECO.2015.01.008.

Ozcan, B., Apergis, N., & Shahbaz, M. (2018). A revisit of the environmental Kuznets curve hypothesis for Turkey: new evidence from bootstrap rolling window causality. *Environmental Science and Pollution Research, 1*–14. https://doi.org/10.1007/s11356-018-3165-x.

Ozturk, I., & Al-Mulali, U. (2015). Investigating the validity of the environmental Kuznets curve hypothesis in Cambodia. *Ecological Indicators, 57*, 324–330. https://doi.org/10.1016/J.ECOLIND.2015.05.018.

Ozturk, I., Al-Mulali, U., & Saboori, B. (2016). Investigating the environmental Kuznets curve hypothesis: the role of tourism and ecological footprint. *Environmental Science and Pollution Research, 23*(2), 1916–1928. https://doi.org/10.1007/s11356-015-5447-x.

Pearce, D., & Barbier, E. (2000). *Blueprint for a Sustainable Economy*. London: Earthscan Publications.

Pennino, M. G., Bellido, J. M., Conesa, D., Coll, M., & Tortosa-Ausina, E. (2017). The analysis of convergence in ecological indicators: An application to the Mediterranean fisheries. *Ecological Indicators, 78*, 449–457. https://doi.org/10.1016/J.ECOLIND.2017.03.041.

Proskuryakova, L. (2018). Updating energy security and environmental policy: Energy security theories revisited. *Journal of Environmental Management, 223*, 203–214. https://doi.org/10.1016/J.JENVMAN.2018.06.016.

Rafiq, S., Salim, R., & Nielsen, I. (2016). Urbanization, openness, emissions, and energy intensity: A study of increasingly urbanized emerging economies. *Energy Economics, 56*, 20–28. https://doi.org/10.1016/J.ENECO.2016.02.007.

Rausser, G., Stevens, R., & Torani, K. (2010). Managing R&D risk in renewable energy: Biofuels vs. alternate technologies. *The Journal of Agrobiotechnology Management&Economics,*

13(4). Retrieved from: http://www.agbioforum.org/v13n4/v13n4a10-torani.htm.

REN21. (2017). *Renewables 2017 global status report.* Retrieved from www.ren21.net/gsr.

Reyes, R. C. (2011). The Green Solow model with natural resources constraint: A theoretical note. *DLSU Business and Economics Review, 21*(1), 111–116.

Rezai, A., & Stagl, S. (2016). *Ecological macroeconomics: Introduction and review* (No. Nr. 9/Year 2/2016). Ecological Economics Papers. Vienna.

Rezai, A., Taylor, L., & Mechler, R. (2013). Ecological macroeconomics: An application to climate change. *Ecological Economics, 85,* 69–76. https://doi.org/10.1016/J.ECOLECON.2012.10.008.

Rogelj, J., den Elzen, M., Höhne, N., Fransen, T., Fekete, H., Winkler, H., et al. (2016). Paris agreement climate proposals need a boost to keep warming well below 2 °C. *Nature, 534*(7609), 631–639. https://doi.org/10.1038/nature18307.

Røpke, I. (2013). Ecological macroeconomics: Implications for the roles of consumer-citizens. In M. J. Cohen, H. S. Brown, & P. J. Vergragt (Eds.), *Innovations in sustainable consumption: New economics, socio-technical transitions and social practices* (pp. 48–64). Cheltenham: Edward Elgar.

Saboori, B., & Sulaiman, J. (2013). Environmental degradation, economic growth and energy consumption: Evidence of the environmental Kuznets curve in Malaysia. *Energy Policy, 60,* 892–905. https://doi.org/10.1016/J.ENPOL.2013.05.099.

Sadorsky, P. (2013). Do urbanization and industrialization affect energy intensity in developing countries? *Energy Economics, 37,* 52–59. https://doi.org/10.1016/J.ENECO.2013.01.009.

Selden, T. M., & Song, D. (1995). Neoclassical growth, the J Curve for abatement, and the inverted U Curve for pollution. *Journal of Environmental Economics and Management, 29*(2), 162–168. https://doi.org/10.1006/jeem.1995.1038.

Shahbaz, M., Solarin, S. A., Hammoudeh, S., & Shahzad, S. J. H. (2017). Bounds testing approach to analyzing the environment Kuznets curve hypothesis with structural beaks: The role of biomass energy consumption in the United States. *Energy Economics, 68,* 548–565. https://doi.org/10.1016/J.ENECO.2017.10.004.

Shahbaz, M., Solarin, S. A., & Ozturk, I. (2016). Environmental Kuznets Curve hypothesis and the role of globalization in selected African countries. *Ecological Indicators, 67,* 623–636. https://doi.org/10.1016/J.ECOLIND.2016.03.024.

Skogen, K., Helland, H., & Kaltenborn, B. (2018). Concern about climate change, biodiversity loss, habitat degradation and landscape change: Embedded in different packages of environmental concern? *Journal for Nature Conservation, 44,* 12–20. https://doi.org/10.1016/J.JNC.2018.06.001.

Solomon, S., Daniel, J. S., Sanford, T. J., Murphy, D. M., Plattner, G.-K., Knutti, R., et al. (2010). Persistence of climate changes due to a range of greenhouse gases. *Proceedings of the National Academy of Sciences of the United States of America, 107*(43), 18354–18359. https://doi.org/10.1073/pnas.1006282107.

Song, T., Zheng, T., & Tong, L. (2008). An empirical test of the environmental Kuznets curve in China: A panel cointegration approach. *China Economic Review, 19*(3), 381–392. https://doi.org/10.1016/J.CHIECO.2007.10.001.

Soytas, U., & Sari, R. (2009). Energy consumption, economic growth, and carbon emissions: Challenges faced by an EU candidate member. *Ecological Economics, 68*(6), 1667–1675. https://doi.org/10.1016/J.ECOLECON.2007.06.014.

Stagl, S. (2014). Ecological macroeconomics: reflections on labour markets. *European Journal of Economics and Economic Policies: Intervention, 11*(2), 171–181. https://doi.org/10.4337/ejeep.2014.02.04.

Stern, D. I. (2015). *The environmental Kuznets curve after 25 years* (CCEP Working Paper No. 1514). Crawford.

Stern, D. I. (2017). The environmental Kuznets curve after 25 years. *Journal of Bioeconomics, 19*(1), 7–28. https://doi.org/10.1007/s10818-017-9243-1.

Stokey, N. L. (1998). Are there limits to growth? *International Economic Review, 39*(1), 1. https://doi.org/10.2307/2527228.

Sulemana, I., James, H. S., & Rikoon, J. S. (2017). Environmental Kuznets Curves for air pollution in African and developed countries: exploring turning point incomes and the role of democracy. *Journal of Environmental Economics and Policy, 6*(2), 134–152. https://doi.org/10.1080/21606544.2016.1231635.

The Lancet Planetary Health. (2018). Can the Paris agreement save us from a climate catastrophe? *The Lancet Planetary Health, 2*(4), e140. https://doi.org/10.1016/S2542-5196(18)30056-1.

Timmons, D., Harris, J. M., & Roach, B. (2014). *The economics of renewable energy.* Retrieved from: http://ase.tufts.edu/gdae.

Ulucak, R. (2018). The evolution of environmental viewpoint in economics: ecological macroeconomics. *Erciyes University Journal of Faculty of Economics and Administrative Sciences, 51,* 127–149. https://doi.org/10.18070/erciyesiibd.402928.

Ulucak, R., & Apergis, N. (2018). Does convergence really matter for the environment? An application based on club convergence and on the ecological footprint concept for the EU countries. *Environmental Science & Policy, 80*(2), 21–27. https://doi.org/10.1016/j.envsci.2017.11.002.

Ulucak, R., & Bilgili, F. (2018). A reinvestigation of EKC model by ecological footprint measurement for high, middle and low income countries. *Journal of Cleaner Production.* https://doi.org/10.1016/j.jclepro.2018.03.191.

UNCED. (1992). *Agenda 21.* Retrieved from: https://sustainabledevelopment.un.org/outcomedocuments/agenda21.

Victor, P. A. (2005). Weak versus strong Sustainability, Eric Neumayer, 2nd edition, Edward Elgar Publishing (2003), 256 pp., ISBN: 184064060X. *Ecological Economics, 52*(1), 127–128. https://doi.org/10.1016/J.ECOLECON.2004.07.006.

Victor, D. G., Akimoto, K., Kaya, Y., Yamaguchi, M., Cullenward, D., & Hepburn, C. (2017). Prove Paris was more than paper promises. *Nature, 548*(7665), 25−27. https://doi.org/10.1038/548025a.

Wackernagel, M., Lin, D., Hanscom, L., Galli, A., & Iha, K. (2019). Ecological footprint. In *Encyclopedia of ecology: Reference module in earth systems and environmental sciences* (pp. 270−282). Elsevier. https://doi.org/10.1016/B978-0-12-409548-9.09567-1.

Wang, F., Ge, Q., Chen, D., Luterbacher, J., Tokarska, K. B., & Hao, Z. (2018). Global and regional climate responses to national-committed emission reductions under the Paris agreement. *Geografiska Annaler − Series A: Physical Geography, 100*(3), 240−253. https://doi.org/10.1080/04353676.2018.1488538.

Wang, X., Jiang, D., & Lang, X. (2017). Future extreme climate changes linked to global warming intensity. *Science Bulletin.* https://doi.org/10.1016/J.SCIB.2017.11.004.

Wang, B., Wang, Q., Wei, Y.-M., & Li, Z.-P. (2018). Role of renewable energy in China's energy security and climate change mitigation: An index decomposition analysis. *Renewable and Sustainable Energy Reviews, 90*, 187−194. https://doi.org/10.1016/J.RSER.2018.03.012.

Wesley Burnett, J., & Madariaga, J. (2017). The convergence of U.S. state-level energy intensity. *Energy Economics, 62*, 357−370. https://doi.org/10.1016/J.ENECO.2016.03.029.

World Commission on Environment and Development. (1987). *Our common future.* Oxford: Oxford University Press. https://doi.org/10.2307/2621529.

Worster, D. (1994). *Nature's economy: A history of ecological ideas.* Cambridge University Press.

Xu, Q., Dong, Y., & Yang, R. (2018). Urbanization impact on carbon emissions in the Pearl River Delta region: Kuznets curve relationships. *Journal of Cleaner Production, 180*, 514−523. https://doi.org/10.1016/J.JCLEPRO.2018.01.194.

Yusuf, A. A. (2004). *Poverty and environmental degradation: Searching for theoretical linkages.*

Zhang, Y.-X., Chao, Q.-C., Zheng, Q.-H., & Huang, L. (2017). The withdrawal of the U.S. from the Paris Agreement and its impact on global climate change governance. *Advances in Climate Change Research, 8*(4), 213−219. https://doi.org/10.1016/J.ACCRE.2017.08.005.

CHAPTER 6

Renewable Energy and its Finance as a Solution to the Environmental Degradation

NICHOLAS APERGIS, PHD

INTRODUCTION

Because energy accounts for the vast majority of greenhouse gas (GHG) emissions, clean energy is the key to addressing climate change. It also brings benefits of energy diversification and security, technology innovation and green jobs, as well as sustainable economic growth and health improvements. Clean energy is at an inflection point as rapid cost declines have facilitated significant growth of the industry. However, the clean energy sector is capital-intensive, with high upfront costs and payback materializing over subsequent years. As such, the ability to mobilize capital and facilitate efficient financing is particularly important. Clean energy companies often look to the capital markets to meet their capital needs, but because of a variety of barriers, there is still insufficient capital available relative to the global need.

Alternative (clean) energy is also increasingly the subject of policy discussion in light of structural changes in global macroconditions, including the following:

- Continued global economic growth: The global economy is witnessing a realignment of economic activity, with Brazil, Russia, India, and China (collectively, the BRIC's economies) leading the way. Continued increases in energy production will be necessary to support this growth.
- Continued population growth: The world population continues to grow, and in 2008—for the first time in history—more than half of the global population, or 3.3 billion people, will be living in urban areas. By 2030, this urban population is expected to grow to almost 5 billion, creating unprecedented energy and land-use concerns.
- Energy security: Political instability in petroleum-rich countries is leading to a broader re-evaluation of energy policy, particularly regarding reliance on imported oil.
- Climate change: Scientific evidence regarding the threats posed by climate change and the extent to which this change is caused by human activity (e.g., GHG emissions associated with energy production and use) is raising important questions about how to achieve sustainable growth and the degree to which alternative sources of energy can substitute for fossil fuels.

Successful policy design, financing, and development of renewable energy (RE) projects require information to guide the decision-making process. Much of this information, however, is not widely available. Project financial terms are especially difficult to come by as many of these terms are often negotiated between two or more private entities. This lack of transparency may also impede effective policy design, competition, and potential industry growth if important information is known by only a few market participants.

Mobilizing finance for investment and innovation in low-carbon energy is a key challenge for climate change mitigation (Dangerman & Schellnhuber, 2013; Grubb, 2014; Stern, 2015). Because cumulative carbon emissions determine the intensity of climate change, speed matters. Yet, fossil fuel investments continue to dwarf investments into RE. In 2015, RE received investments of less than USD 260 billion, which represented only 16% of the USD 1.6 trillion in total energy sector investments. Meanwhile, investment in fossil fuels in the power sector, where they compete directly with electricity from RE, rose by 7% from 2013 to 2014 (UNEP & BNEF, 2015). Clearly, fossil fuels still dominate energy investment; therefore, a major concern in the transition to low-

Environmental Kuznets Curve (EKC) https://doi.org/10.1016/B978-0-12-816797-7.00006-0

carbon energy provision is how to obtain enough finance to steer investments into the RE direction. Total funding for RE, however, has been rising at a remarkable rate. According to Bloomberg New Energy Finance (BNEF), the amount of RE finance, along the entire innovation chain, from research and development (R&D) for new technologies to asset finance for full-scale power plants, rose from USD 45 billion in 2004 to 270 billion in 2014 globally. This represents a compound annual growth rate of 18%. Moreover, in 2014, net investment into new capacity, as opposed to replacing depreciated assets, was twice as large for RE as it was for fossil fuels in the power sector; this trend is forecast to continue for the rest of this decade (International Energy Agency, 2015). Therefore, although investment in RE remains low relative to that in fossil fuels, the trajectory is a positive one. The focus on achieving a greater amount of finance has diverted attention from what is being financed. Because finance flows toward concrete projects and firms, finance always creates a direction toward areas and technologies that these organizations promote. This may result in a skewed distribution of investment in RE, so that some areas are overfinanced, whereas others are underfinanced. Lack of attention on the relationship between finance and directionality is surprising because it is widely recognized that a diverse set of RE technologies is desirable for at least two reasons: (1) with a wide portfolio, if innovation is unsuccessful in one area, not all eggs are in one basket (Grubler, 2012) and (2) a diversified energy supply increases resilience of the energy system and hence energy security (Stern, 2015; Stirling, 2010b).

There has been much research linking the research and commercialization phase of the innovation chain to specific financing needs. High-risk upstream research is widely understood to require public financing because of the characteristics of public goods (Arrow, 1962). Similarly, venture capital financing helps to solve the asymmetric information problem in the "Valley of Death," which requires carrying technologies from proof of concept to commercial scale (Auerswald & Branscomb, 2003). However, less studied are the diverse types of finance in the downstream phase of innovation: deployment and diffusion. And yet, more than two-thirds of total RE finance went to asset finance for deployment of utility-scale RE power plants and so can affect directions in innovation. Channels of influence work both directly through the finance committed favoring a certain technology and indirectly through the effects of increasing returns to scale and learning by doing, where feedback loops from deployment to upstream innovation can create technology lock-ins (Arthur, 1989). Yet the literature on the "directionality" of innovation, which has looked for example at the way that policy measures can affect directions of innovation either knowingly or unknowingly (Stirling, 2010a), has ignored the role of finance in this process.

FINANCE AND ENERGY INNOVATION
Financial Actors and Innovation Directions

Joseph Schumpeter placed finance at the center of his theory of innovation, as providing the funds necessary for the entrepreneur to spring into action. However, he focused on only one type of finance: banks (Schumpeter, 1939) and did not elaborate on the question of whether different financial actors' characteristics might impact what innovation is being financed, thus, creating directions. The Miller-Modigliani theorem, which states that sources of finance (equity or debt financing from any actor) do not matter to firms and hence do not affect the real economy (Modigliani & Miller, 1959) has further detracted attention away from distinguishing between types of finance in innovation. In subsequent literature, the only types of actors typically singled out were "government" and "venture capitalists" (Hall, 2002). The job of the former was to overcome underinvestment in research due to the positive externality of knowledge (Arrow, 1962); the purpose of the latter was to overcome information asymmetries that led to underinvestment into product development by new firms or "ventures" (Hall & Lerner, 2009). In this literature, finance takes a passive role regarding what is being financed. More recent work has placed greater emphasis on different types of financial actors and how they may impact the characteristics of the firms and technologies they are financing. Thus, financing by the public sector also beyond the R&D stage (Mazzucato, 2013) in areas such as space, health, and low-carbon technology has resulted in the creation of whole new sectors, often through mission-oriented projects that were actively decided upon by those who provided the finance (Foray, Mowery, & Nelson, 2012). In the private sector, certain actors were also pushing particular sectors or technologies (Mazzucato & Wray, 2015). But what gets financed may equally be influenced by what is neglected by certain actors: it has been noted that venture capital has often avoided very early-seed investments and has also been biased toward particular areas such as IT and biotech, only recently getting interested in green-tech (Lerner,

2012). Some studies have examined how short-term speculative financial actors have affected science-based industries (Lazonick & Tulum, 2011; Pisano, 2006).

Conversely, the literature concerned with directions has paid little attention to the role of finance in setting these directions. The directionality literature (Stirling, 2010a, 2011) in innovation studies has stressed the importance of recognizing the multiple pathways and directions that innovation can take, so that policies explicitly recognize the forces influencing them, including the risk of suboptimal policies and lock-ins. This strand of literature has focused on the role of power relations, such as those embodied in public financing of innovation. However, it has ignored how the distribution and characteristics of private and public financial actors can affect the direction of change. Similarly, economic studies considering path dependence in innovation (David, 1985) and the role of feedback effects in creating "lock-in" (Arthur, 1989) have not included the way that financial institutions can affect this dynamic.

Financial Actors and Direction in Renewable Energy

The literature on RE financing, both modeling and empirical, has historically given more attention to sufficient investment in R&D than to downstream financing of deployment (Popp, 2011; Sagar & van der Zwaan, 2006). Yet, a key gap identified in RE more recently is the lack of finance for downstream capital-intensive high-risk projects (European Commission, 2013; Veugelers, 2012; Zindler & Locklin, 2010), kindling a growing literature that studies actors in the deployment of RE technologies. One strand focuses on different risk appetites of types of finance. Ghosh and Nanda (2010) have argued that the capital required for asset finance of the capital-intensive RE power plants is typically an order of magnitude larger than that which venture capitalists have been willing to supply for technology development (Gaddy, Sivaram, & OSullivan, 2016) and too risky for banks (Kalamova, Kaminker, & Johnstone, 2011). Evidence for differences in risk perception between individual investors has been furnished for a sample of European investors by Masini and Menichetti (2012). Although these studies do not distinguish types of financial actors, Bergek, Mignon, and Sundberg (2013) study three types of power plant builders in Sweden—utilities, farmers, cooperatives—and highlight how different builder types may have various nonprofit maximization objectives that influence their investment choices. The results connect

with conceptual work by Langniss (1996), who identifies six financial actors (anonymous, industry, large utility, house owner, municipality, and energy community) and discusses how each type's appetite for risk varies with their motives for investment.

Another strand focuses on the impact of public policies on private deployment finance. Studies that used aggregate data to examine the impact of innovation policies by government aimed at upstream innovation on private RE deployment (Johnstone, Haščič, & Popp, 2009; Popp, Haščič, & Medhi, 2011), and at private downstream activities (Eyraud, Clements, & Wane, 2013) find that these policies do mobilize private finance. Using microdata at the asset deal level (mostly from BNEF, as in the present study), public policies are found to mobilize finance from institutional investors (Polzin, Migendt, Täube, & von Flotow, 2015) and to have a positive effect on cross-border merger and acquisition activity (Criscuolo, Johnstone, Menon, & Shestalova, 2014). Certain types of policies are more conducive to investment in RE innovation than others (Veugelers, 2012) and may induce varying amounts of venture capital investments into RE companies (Criscuolo & Menon, 2015). Only two studies have distinguished direct public investments (Cárdenas Rodríguez, Haščič, Johnstone, Silva, & Ferey, 2014; Haščič, Cárdenas Rodríguez, Jachnik, Silva, & Johnstone, 2015). They found that both public investments and policies have a significant positive impact on private investment. In addition, Cárdenas Rodríguez et al. (2014) show that direct public investments are taking place for those technologies, where other public policies have had little effect on mobilizing private finance. As for directions, the directionality literature has considered the energy sector, but focus on the interplay of agency and structure and the influence of power without distinguishing finance (Stirling, 2014). Meanwhile in economics, discussion of the direction of technical change in energy (distinguishing fossil and RE directions) incorporates an influence of financial actors neither in the phase of R&D (Acemoglu, Akcigit, Hanley, & Kerr, 2016) nor in that of deployment (Jaffe, Newell, & Stavins, 2005).

Finance and the Role of Feed-In Tariffs

The current cost disadvantages of alternative technologies that make it difficult for them to compete with fossil fuel–sourced energy can be offset through various policy mechanisms. These policies encourage (or even mandate) the use of energy produced by alternative technologies in the hopes of eventually lowering the costs of and establishing a predictable investment environment for such technologies. The most popular of those mechanisms comes from the feed-in tariffs.

A feed-in tariff requires a utility to purchase power from an RE source at a fixed rate, typically above the market price. By establishing a set price, feed-in tariffs can generate stability and reduce financing barriers and costs for alternative energy sources. A typical program distributes the overall cost of the subsidies (i.e., above market price) to ratepayers through a fee on electricity sales. Feed-in tariff programs typically include stable long-term payments (10-year minimums), a scheduled decrease in the subsidy over time (known as a tariff digression), varying incentive levels for each different technology, and stepped tariffs, which target specific types of technology to encourage a more efficient use of energy. There are two primary types of feed-in tariffs: the standard offer and the advanced renewable. The standard offer feed-in tariff sets the same price across all RE technologies and project sizes. The advanced renewable feed-in tariff offers a distinct price for each different technology, and rates fluctuate to ensure profitable operation of RE projects after installation. The rates in the advanced renewable approach range dramatically among technologies and also fluctuate throughout the day. Decreasing the subsidy over time creates a predictable investment environment while producing pressure on the RE industry to reduce the costs of a technology and has been shown to increase competition among manufacturers.

The literature on the role of policy in the development of RE sources is vast. However, the majority of research takes a normative or descriptive approach to outlining the factors that influence RE development. Several case studies (del Río González and Gual, 2007; del Río González, 2008; Haas et al., 2011; Lesser & Su, 2008; Lipp, 2007) and other qualitative evaluation techniques have suggested that feed-in tariff policies are an important element in explaining the success of RE development in Europe. Rigorous empirical studies of RE policy effectiveness are less common. Most econometric studies assessing the effectiveness of RE policies to date have focused on state-level policies in the United States. Studies that examine the drivers of RE development can be divided into three groups: (1) those that use cross-sectional specifications, (2) those that use panel data to control for state-level characteristics, and (3) more nuanced analyses that use complex dependent and/or independent variables to articulate differences in policy design or policy responsiveness.

The first group (Adelaja & Hailu, 2008; Menz & Vachon, 2006) uses pooled cross-section regressions to look at the impact of policy variables (usually a binary variable or simple numeric indicator such as the dollar value of a tax credit) on RE capacity development. Alegappan et al. (2011) rely on descriptive statistics only. These studies find a strong positive correlation between RE policy and RE development. However, their specifications do not control for state-level characteristics or time trends that may be correlated with both policy implementation and RE development, so this relationship cannot be interpreted as causal. The second group (Carley, 2009; Delmas & Montes-Sancho, 2011; Marques, Fuinhas, & Pires Manso, 2010; Shrimali & Kneifel, 2011) provides empirical evidence to support this concern. These studies use fixed-effects regression models or other strategies to reduce omitted variables bias from state characteristics that are correlated with both policy implementation and RE deployment. They suggest a less certain relationship between RE policies and deployment. For example, Carley (2009) finds that RE policies implementation does not predict the percentage of energy generation from renewable sources, although the number of years a state maintains an RE policy is a significant determinant of total RE capacity development. Shrimali and Kneifel (2011) find that RE policies actually appear to reduce the penetration of some RE technologies and overall RE capacity, while they increase the penetration of others. These results call into question the effectiveness of RE policies once nonpolicy state characteristics have been controlled for. The third group (Marques, Fuinhas, & Manso, 2011; Yin & Powers, 2009) uses more nuanced model specifications to better capture the complexity of RE development. Marques et al. (2011) apply a quantile regression approach to analyze the drivers of RE deployment in EU countries, finding that responsiveness to economic and social drivers varies in magnitude, significance, and sometimes direction between countries with different initial levels of RE penetration. However, Marques et al. (2011) do not incorporate any policy variables. Yin and Powers (2009) make a key contribution to the debate by addressing policy design heterogeneity in state-level RE policies in the United States. They develop a new quantitative measure of RE stringency that takes into account policy design features that differ by state. Applying this new measure within a fixed-effect specification, they find that RE policies have had a significant and positive effect on RE deployment. Most importantly, they verify that this effect would not be observed if differences in policy design are ignored, as done in studies that use binary policy variables only.

A different strand of the relevant literature deals with the mechanisms of financing clean energy diffusion and infrastructure. In this term, certain sources of finance are available as certain problems, such as technologic (and market) uncertainty and lack of collateral are overcome. However, technology-specific problems, such as long payback periods and policy uncertainty still prevail. The remaining regulatory uncertainty implies the default risk on any project is substantial and importantly, not calculable. Therefore, credit ratings are low or absent and only (private) equity or junk bonds can channel savings into these types of investments. Corporate finance mostly comes from insurances, banks, or pension funds (Frankfurt School-UNEP Center, 2016). They finance larger, mature clean energy companies, as well as projects and infrastructure with debt or equity. Debt investors do not receive dividends and do not benefit from higher profitability and cost reductions directly and hence care about the downside much more than they do about the potential upsides. Faced with such opposition, the incentives for debtors are to reduce leverage to have a better credit rating and, thus, lower cost of capital, increasing the value of the stock in the short run. The focus of the ecosystem for financing toward debt generates a bias toward calculable risks in incremental innovation and maintenance of the capital stock in existing firms. They have high-quality marketable collateral and established track records but lack the incentive to introduce and diffuse true innovations as they cannibalize on existing profit flows (Hockerts & Wüstenhagen, 2010).

Based on the above discussion, it is worth mentioning certain policies to stimulate investments into clean energy. Central governments arguably have the means to break the "lock-in" problems, which favor fossil fuel–based energy technologies (Smink, Hekkert, & Negro, 2015). Mistakes, however, are terribly costly and can generate new lock-ins for which politicians do not want to be held accountable. Sticking to the existing system may be more attractive, even at the country level, than running the chance of locking into a losing technology. The obvious angle from which policy-makers could approach the challenges for the energy transition is innovation policy (van den Bergh, 2013). Market-based incentives such as carbon emission trading systems represent the theoretic optimum as argued by climate and energy economics since the early 1990s (Fischer & Newell, 2008; Polzin et al., 2015). However, due to the lack of global mechanisms, second-best instruments are required. To accelerate the diffusion of clean energy and associated investments, policy-makers first could deploy technology-push mechanisms such as direct R&D investments, subsidies,

and tax-credits that target the early stages of the innovation cycle (Bergek et al., 2013; Bürer & Wüstenhagen, 2009; Olmos, Ruester, & Liong, 2012). Direct investments and cofunding also mobilize private early-stage finance (Bird et al., 2005; Menz & Vachon, 2006). Olmos et al. (2012) suggest public loans, or guarantees provided by public bodies backing private loans, along with public investments in the equity of innovating companies to accelerate the commercialization. Second, research on clean energy diffusion and investment highlights demand-pull policies mostly targeting the later stages of the innovation cycle (Bürer & Wüstenhagen, 2009; Haley & Schuler, 2011; Hoppmann, Peters, Schneider, & Hoffmann, 2013). Fiscal and financial incentives, such as grants and subsidies (Bergek et al., 2013; Olmos et al., 2012; Roy, Ghosh, Ghosh, & Dasgupta, 2013) prove less effective than feed-in tariffs (Couture & Gagnon, 2010; del Río & Bleda, 2012; Marques & Fuinhas, 2012; Jenner, Groba, & Indvik, 2013; Rodríguez, Haščič, Johnstone, Silva, & Ferey, 2015) that also target smaller distributed capacity investments (Vasileiadou, Huijben, & Raven, 2016). To address network externalities and reduce private risks regarding complementary assets, public investments are suggested (Henriot, 2013; Steinbach, 2013). The (quality) regulation of the (clean energy) portfolio and emission standards advances deployment of more mature technologies (Carley, 2009; Delmas & Montes-Sancho, 2011). Systemic policies such as long-term planning and policy support accelerate both early and later investments (Bergek et al., 2013; Masini & Menichetti, 2012; Wüstenhagen & Menichetti, 2012). Especially consistency, stringency, and predictability to reduce deep uncertainty and policy risk are deemed crucial (Lüthi & Wüstenhagen, 2012; White, Lunnan, Nybakk, & Kulisic, 2013). Overall, a policy mix is suggested to make the transition (Rogge & Reichardt, 2016). Most of the policies actually favor mature established technologies, appealing to the debt-based financial markets. In the case of the Eurozone, unprecedented monetary policies, such as the Quantitative Easing mechanism, have driven the cost of debt finance to zero or below and flooded financial markets. But only very little of that monetary expansion finds its way into the real economy, let alone into clean energy (Frankfurt School-UNEP Center, 2016). Instead, these policies tend to entrench the existing linear, carbon-based economy, as debt favors low-risk status quo. More equity is needed to finance high-uncertainty innovation. Hence, the second angle from which policy-makers could mobilize private investment for clean energy innovation, and diffusion revolves around framework conditions and regulation of financial markets. First, framework conditions for either debt- or equity-based instruments influence their contribution to a clean energy transition, as a developed capital market is needed to channel resources (Cumming, Henriques, & Sadorsky, 2016; Da Rin, Nicodano, & Sembenelli, 2006). Most importantly, a fiscal preferential treatment of debt finance should be avoided. Interest is deductible as costs, while dividend payments only occur after tax. Hence a favorable tax policy could allow for tax deductibility of early-stage company investments instead (Keuschnigg & Nielsen, 2003; Lerner & Tåg, 2013). A less stringent bankruptcy and labor market legislation creates entrepreneurship and experimentation in sectors characterized by high uncertainty, such as clean-tech (Cumming, 2011). Second, capital market regulation shapes investment mandates, risk models and assessment, and capital market rules and thus ultimately determines the feasibility and viability of investments into clean energy (Schmidt, 2014). Regulation (such as Basel III), especially since the crisis, is almost exclusively geared toward stability and security (Lerner & Tåg, 2013). Consequently, they encourage or force deposits into "safe" asset classes and calculable risks, such as rated firms, government debt and real estate. Institutional investors and their intermediaries are forced to stay away from risky asset classes (Block & Sandner, 2009; Marcus, Malen, & Ellis, 2013). New alternative finance, such as equity- and debt-based crowdfunding is also becoming more regulated in many countries (Vismara, 2016). Regulators abstain from clamping down on shadow banking and these new forms of intermediation. It is more promising to be clear about the fact that such investments are not regulated and that investors willingly accept uncertainty and risk than to try and protect investors. But the volumes to be expected from these emerging intermediation channels are limited. Although many clean energy investments projects are economical, the question is why even these are not funded under record low-interest rates. This problem stems from unintended consequences of stability-oriented regulation toward actors most favorable of long-term RE projects, thereby stifling monetary easing and making it virtually ineffective.

Conclusion

Public policy-makers, industry, rate-paying customers, and other stakeholders on a global basis are increasingly viewing RE as an important component of strategies aimed at addressing the challenges associated with energy production and use. The principal challenges include the trade-off between economic growth, environmental impacts, and national security. Driven by these combined pressures, a number of energy policies put in place around the world during the last decade have helped the RE industry to grow significantly. Because most RE technologies are not yet, in most regions, able to compete economically with fossil fuels, they will have to be supported by public policy interventions if RE is to play a real near-term role in energy policy. Public policies are currently necessary to reduce the costs and improve the investment environment to enable significant and long-term growth in the use of RE.

As illustrated by this survey, there is no single or best solution to the challenge of encouraging the use of RE. Rather, picking the most effective options will depend on the specific conditions of each region and market, as well as the particular characteristics of each type of RE. There are, however, a range of available policy options that have been tested in various places around the world. That base of experience should prove helpful for policy-makers who wish to expand the use of RE.

At the same time, this survey also touched the discussion about mobilizing private finance for clean energy innovation. The literature has neglected the structure and regulation of financial markets as potential determinants. To mobilize resources to break out of the fossil fuel technology complex and to finance radical and transformative innovations, the industry needs intermediation to take different channels. Although low-risk debt is suitable to finance diffusion, the industry needs a shift to more expensive and uncertain equity to finance innovation. However, financial market regulations are currently "boxing in" incumbents, which are not responding to the financial requirements of an innovation-led energy transition, such as early-stage risk capital. Financial regulatory reforms could free up the resources in banks and institutional investors for more uncertain and equity-like intermediation. Regulators could also push for such higher equity ratios in banks and intermediaries themselves to justify taking higher risk. Implicit and explicit guarantees for deposits and other debt liabilities on the banks' balance sheets must be eliminated. This sets the necessary preconditions for a more diverse financial sector in which all varieties of intermediation can compete on a level playing field and implicit public support for banks no longer tilts the system toward cheap debt finance of status quo assets and allows for a transition toward clean energy.

REFERENCES

Acemoglu, D., Akcigit, U., Hanley, D., & Kerr, W. (2016). Transition to clean technology. *Journal of Political Economy*, *124*, 52–104.

Adelaja, S., & Hailu, Y. G. (2008). *Effects of renewable portfolio standards and other state policies on wind industry development in the*. East Lansing, MI: U.S. Michigan State University.

Alagappan, L., Orans, R., & Woo, C. (2011). What drives renewable energy development? *Energy Policy*, *39*, 5099–5104.

Arrow, K. J. (1962). The economic implications of learning by doing. *The Review of Economic Studies*, *29*, 155–173.

Arthur, W. B. (1989). Competing technologies, increasing returns, and lock-in by historical events. *Economic Journal*, *99*, 116–117.

Auerswald, P. E., & Branscomb, L. M. (2003). Valleys of death and Darwinian seas: Financing the invention to innovation transition in the United States. *The Journal of Technology Transfer*, *28*, 227–239.

Bergek, A., Mignon, I., & Sundberg, G. (2013). Who invests in renewable electricity production? Empirical evidence and suggestions for further research. *Energy Policy*, *56*, 568–581.

van den Bergh, J. C. J. M. (2013). Environmental and climate innovation: Limitations, policies and prices. *Technological Forecasting and Social Change*, *80*, 11–23.

Bird, L. A., Bolinger, M., Gagliano, T., Wiser, R., Brown, M., & Parsons, B. (2005). Policies and market factors driving wind power development in the United States. *Energy Policy*, *33*, 1397–1407.

Block, J., & Sandner, P. (2009). What is the effect of the financial crisis on venture capital financing? Empirical evidence from US Internet start-ups. *Venture Capital International Journal of Entrepreneurship Finance*, *11*, 295–309.

Bürer, M. J., & Wüstenhagen, R. (2009). Which renewable energy policy is a venture capitalist's best friend? Empirical evidence from a survey of international clean-tech investors. *Energy Policy*, *37*, 4997–5006.

Cárdenas Rodríguez, M., Haščič, I., Johnstone, N., Silva, J., & Ferey, A. (2014). *Inducing private finance for renewable energy projects: Evidence from micro-data*. OECD Environment Working Paper, No. 67.

Carley, S. (2009). State renewable energy electricity policies: An empirical evaluation of effectiveness. *Energy Policy*, *38*, 3071–3081.

Couture, T., & Gagnon, Y. (2010). An analysis of feed-in tariff remuneration models: Implications for renewable energy investment. *Energy Policy*, *38*, 955–965.

Criscuolo, C., Johnstone, N., Menon, C., & Shestalova, V. (2014). *Renewable energy policies and cross-border investment: Evidence from mergers and acquisitions in solar and wind energy*. OECD Science, Technology and Industry Working Paper, No. 2014/03.

Criscuolo, C., & Menon, C. (2015). Environmental policies and risk finance in the green sector: Cross-country evidence. *Energy Policy*, *83*, 38–56.

Cumming, D. (2011). Public policy and the creation of active venture capital markets. *Venture Capital*, *13*, 75–94.

Cumming, D., Henriques, I., & Sadorsky, P. (2016). "Cleantech" venture capital around the world. *International Review of Financial Analysis*, *44*, 86–97.

Da Rin, M., Nicodano, G., & Sembenelli, A. (2006). Public policy and the creation of active venture capital markets. *Journal of Public Economics*, *90*, 1699–1723.

Dangerman, A. T. C. J., & Schellnhuber, H. J. (2013). Energy systems transformation. *Procedure of National. Academic Sciences*, *110*, E549–E558.

David, P. A. (1985). Clio and the economics of QWERTY. *The American Economic Review*, *75*, 332–337.

Delmas, M. A., & Montes-Sancho, M. J. (2011). U.S. State policies for renewable energy: Context and effectiveness. *Energy Policy*, *39*, 2273–2288.

European Commission. (2013). Technology assessment. In *Commission staff working document No. SWD(2013) 158 final*.

Eyraud, L., Clements, B., & Wane, A. (2013). Green investment: Trends and determinants. *Energy Policy*, *60*, 852–865.

Fischer, C., & Newell, R. G. (2008). Environmental and technology policies for climate mitigation. *Journal of Environmental Economics and Management*, *55*, 142–162.

Foray, D., Mowery, D. C., & Nelson, R. R. (2012). Public R&D and social challenges: What lessons from mission R & D programs? *Resources Policy*, *41*, 1697–1702.

Frankfurt School-UNEP Centre, BNEF. (2016). *Global trends in renewable energy investment*. Working Paper.

Gaddy, B., Sivaram, V., & OSullivan, F. (2016). *Venture capital and clean tech: The wrong model for clean energy innovation*. MITEI-Working Paper, No. 2016-06.

Ghosh, S., & Nanda, R. (2010). *Venture capital investment in the clean energy sector*. Harvard Business School Working Paper, No. 11-020.

Grubb, M. (2014). *Planetary economics*. Oxford and New York: Routledge.

Grubler, A. (2012). Energy transitions research: Insights and cautionary tales. *Energy Policy*, *50*, 8–16.

Haas, R., Ragwitz, M., Resch, G., Panzer, C., Busch, S., & Held, A. (2011). Efficiency and effectiveness of promotion systems for electricity generation from renewable energy sources – lessons from EU countries. *Energy*, *36*, 2186–2193.

Haley, U. C. V., & Schuler, D. A. (2011). Government policy and firm strategy in the solar photovoltaic industry. *California Management Review*, *54*, 17–38.

Hall, B. H. (2002). The financing of research and development. *Oxford Review of Economic Policy*, *18*, 35–51.

Hall, B. H., & Lerner, J. (2009). *The financing of R&D and innovation*. NBER Working Paper, No. 15325.

Haščič, I., Cárdenas Rodríguez, M., Jachnik, R., Silva, J., & Johnstone, N. (2015). *Public interventions and private climate finance flows: Empirical evidence from renewable energy financing*. OECD Environment Working Paper, No. 80.

Henriot, A. (2013). Financing investment in the European electricity transmission network: Consequences on long-term sustainability of the TSOs financial structure. *Energy Policy*, *62*, 821–829.

Hockerts, K., & Wüstenhagen, R. (2010). Greening Goliaths versus emerging Davids-theorizing about the role of incumbents and new entrants in sustainable entrepreneurship. *Journal of Business Ventures, 25*, 481−492.

Hoppmann, J., Peters, M., Schneider, M., & Hoffmann, V. H. (2013). The two faces of market support-how deployment policies affect technological exploration and exploitation in the solar photovoltaic industry. *Resources Policy, 42*, 989−1003.

International Energy Agency, (2015). *World Energy Outlook 2015.* OECD, France.

Jaffe, A. B., Newell, R. G., & Stavins, R. N. (2005). A tale of two market failures: Technology and environmental policy. *Ecological Economics, 54*, 164−174.

Jenner, S., Groba, F., & Indvik, J. (2013). Assessing the strength and effectiveness of renewable electricity feed-in tariffs in European Union countries. *Energy Policy, 52*, 385−401.

Johnstone, N., Haščič, I., & Popp, D. (2009). Renewable energy policies and technological innovation: Evidence based on patent counts. *Environmental and Resource Economics, 45*, 133−155.

Kalamova, M., Kaminker, C., & Johnstone, N. (2011). *Sources of finance, investment policies and plant entry in the renewable energy sector.* OECD Environment Working Paper, No. 37.

Keuschnigg, C., & Nielsen, S. B. (2003). Tax policy, venture capital, and entrepreneurship. *Journal of Public Economics, 87*, 175−203.

Langniss, O. (1996). Instruments to foster renewable energy investments in Europe: A survey under the financial point of view. *Renewable Energy, 9*, 1112−1115.

Lazonick, W., & Tulum, Ö. (2011). US biopharmaceutical finance and the sustainability of the biotech business model. *Resources Policy, 40*, 1170−1187.

Lerner, J. (2012). The narrowing ambitions of venture capital. *MIT Technology Review, 115*, 76−78.

Lerner, J., & Tåg, J. (2013). Institutions and venture capital. *Industrial and Corporate Change, 22*, 153−182.

Lesser, J. A., & Su, X. (2008). Design of an economically efficient feed-in tariff structure for renewable energy development. *Energy Policy, 36*, 981−990.

Lipp, J. (2007). Lessons for effective renewable electricity policy from Denmark, Germany and the United Kingdom. *Energy Policy, 35*, 5481−5495.

Lüthi, S., & Wüstenhagen, R. (2012). The price of policy risk-empirical insights from choice experiments with European photovoltaic project developers. *Energy Economics, 34*, 1001−1011.

Marcus, A., Malen, J., & Ellis, S. (2013). The promise and pitfalls of venture capital as an asset class for clean energy investment research questions for organization and natural environment scholars. *Organization & Environment, 26*, 31−60.

Marques, A. C., & Fuinhas, J. A. (2012). Are public policies towards renewable successful? Evidence from European countries. *Renewable Energy, 44*, 109−118.

Marques, A. C., Fuinhas, J. A., & Manso, J. (2011). A quantile approach to identify factors promoting renewable energy in European countries. *Environmental and Resource Economics, 49*, 351−366.

Marques, A. C., Fuinhas, J. A., & Pires Manso, J. (2010). Motivations driving renewable energy in European countries: A panel data approach. *Energy Policy, 38*, 6877−6885.

Masini, A., & Menichetti, E. (2012). The impact of behavioural factors in the renewable energy investment decision making process: Conceptual framework and empirical findings. *Energy Policy, 40*, 28−38.

Mazzucato, M. (2013). *The entrepreneurial state.* London and New York: Anthem Press.

Mazzucato, M., & Wray, R. L. (2015). *Financing the capital development of the economy: A Keynes-Schumpeter-Minsky synthesis.* Levy Institute Working Paper, No. 837.

Menz, F. C., & Vachon, S. (2006). The effectiveness of different policy regimes for promoting wind power: Experiences from the states. *Energy Policy, 34*, 1786−1796.

Modigliani, F., & Miller, M. H. (1959). The cost of capital, corporation finance, and the theory of investment: Reply. *The American Economic Review, 49*, 655−669.

Olmos, L., Ruester, S., & Liong, S.-J. (2012). On the selection of financing instruments to push the development of new technologies: Application to clean energy technologies. *Energy Policy, 43*, 252−266.

Pisano, G. P. (2006). *Science business: The promise, the reality and the future of biotech.* Boston: Harvard Business School Press.

Polzin, F., Migendt, M., Täube, F. A., & von Flotow, P. (2015). Public policy influence on renewable energy investments—a panel data study across OECD countries. *Energy Policy, 80*, 98−111.

Popp, D. (2011). *Innovation and climate policy.* NBER Working Paper, No. 15673.

Popp, D., Haščič, I., & Medhi, N. (2011). Technology and the diffusion of renewable energy. *Energy Economics, 33*, 648−662.

del Río González, P. (2008). Ten years of renewable electricity policies in Spain: An analysis of successive feed-in tariff reforms. *Energy Policy, 36*, 2917−2929.

del Río, González, P., & Gual, M. A. (2007). An integrated assessment of the feed-in tariff system in Spain. *Energy Policy, 35*, 994−1012.

del Río, P., & Bleda, M. (2012). Comparing the innovation effects of support schemes for renewable electricity technologies: A function of innovation approach. *Energy Policy, 50*, 272−282.

Rodríguez, M. C., Haščič, I., Johnstone, N., Silva, J., & Ferey, A. (2015). Renewable energy policies and private sector investment: Evidence from financial microdata. *Environmental and Resource Economics, 62*, 163−188.

Rogge, K. S., & Reichardt, K. (2016). Policy mixes for sustainability transitions: An extended concept and framework for analysis. *Resources Policy, 45*, 1620−1635.

Roy, J., Ghosh, D., Ghosh, A., & Dasgupta, S. (2013). Fiscal instruments: Crucial role in financing low carbon transition in energy systems. *Current Opinion of Environmental Sustainability, 5,* 261–269.

Sagar, A. D., & van der Zwaan, B. (2006). Technological innovation in the energy sector: R&D, deployment, and learning-by-doing. *Energy Policy, 34,* 2601–2608.

Schmidt, T. S. (2014). Low-carbon investment risks and derisking. *National Climate Change, 4,* 237–239.

Schumpeter, J. A. (1939). *Business cycles* (Vol. 1). New York and London: McGraw-Hill.

Shrimali, G., & Kneifel, J. (2011). Are government policies effective in promoting deployment of renewable electricity resources? *Energy Policy, 39,* 4726–4741.

Smink, M. M., Hekkert, M. P., & Negro, S. O. (2015). Keeping sustainable innovation on a leash? Exploring incumbents' institutional strategies. *Business Strategy and the Environment, 24,* 86–101.

Steinbach, A. (2013). Barriers and solutions for expansion of electricity grids-the German experience. *Energy Policy, 63,* 224–229.

Stern, N. (2015). *Why are we waiting?: The logic, urgency, and promise of tackling climate change.* MIT Press.

Stirling, A. (2010a). Keep it complex. *Nature, 468,* 1029–1031.

Stirling, A. (2010b). Multicriteria diversity analysis: A novel heuristic framework for appraising energy portfolios. *Energy Policy, 38,* 1622–1634.

Stirling, A. (2011). Pluralising progress: From integrative transitions to transformative diversity. *Environmental Innovation of Social Transition, 1,* 82–88.

Stirling, A. (2014). Transforming power: Social science and the politics of energy choices. *Energy Resources of Social Sciences, 1,* 83–95.

UNEP & BNEF. (2015). *Global trends in renewable energy investment 2015.* Frankfurt School of Finance & Management.

Vasileiadou, E., Huijben, J. C. C. M., & Raven, R. P. J. M. (2016). Three is a crowd? Exploring the potential of crowd-funding for renewable energy in The Netherlands. *Journal of Cleaner Production, 128,* 142–155.

Veugelers, R. (2012). Which policy instruments to induce clean innovating? *Resources Policy, 41,* 1770–1778.

Vismara, S. (2016). Equity retention and the social network theory in equity crowd-funding. *Small Business Economics, 46,* 579–590.

White, W., Lunnan, A., Nybakk, E., & Kulisic, B. (2013). The role of governments in renewable energy: The importance of policy consistency. *Biomass and Bioenergy, 57,* 97–105.

Wüstenhagen, R., & Menichetti, E. (2012). Strategic choices for renewable energy investment: Conceptual framework and opportunities for further research. *Energy Policy, 40,* 1–10.

Yin, H., & Powers, N. (2009). Do state renewable portfolio standards promote in-state renewable generation? *Energy Policy, 38,* 1140–1149.

Zindler, E., & Locklin, K. (2010). *Crossing the Valley of Death.* Bloomberg New Energy Finance.

CHAPTER 7

Data Selection and Environmental Kuznets Curve Models - Environmental Kuznets Curve Models, Data Choice, Data Sources, Missing Data, Balanced and Unbalanced Panels

AVIK SINHA, FPM • MUHAMMAD SHAHBAZ, PHD • DANIEL BALSALOBRE, PHD

INTRODUCTION

The environmental Kuznets curve (EKC) hypothesis is a well-established hypothesis in the literature of energy and environmental economics. While assessing the relationship between economic growth and environmental degradation, Grossman and Krueger (1991) found its resemblance with the findings of Simon Kuznets (1955), and they named this association as the EKC hypothesis. Since then, various studies have been published on various contexts and for various pollutants, either by supporting or by disproving this hypothesis.[1]

However, in the course of EKC analysis, a number of problems are encountered by the researchers. Some of those problems might be categorized as (a) model selection, (b) model validation, (c) data standardization, (d) variable selection and creation, and (e) proxy for environmental degradation. Now, we will look into these problem areas individually. Let us begin with the problem of model selection. While designing an EKC model for any context, a researcher will have to choose between linear, quadratic, and cubic specifications of EKC. Out of these options, it is needed to analyze the specification(s), which might complement the context. To this aim, the theoretic underpinning of all the three models needs to be understood. Next, we will look at the validation of an EKC model. In general, it has been seen that the EKC models derived by several researchers are not mathematically valid. Therefore, we need to devise a set of necessary and sufficient validation conditions for the EKC models, so that the correctness of the models can be checked. Subsequent to that, we will look into the issues of data standardization. Now, this issue can come in multiple forms. First, the data available for the model variables might pertain to different base years. Before proceeding with the data analysis, the base year for all the variables should be standardized. Second, missing data has been a common problem for the EKC studies, as it might not be possible for the data collection agencies to collect and publish the data periodically for all the geographic locations and across all time periods. Sometimes, the different government bodies of any country might choose not to bring some data in the public domain. Owing to these reasons, researchers find missing data points for several variables, while carrying out any EKC analysis. Before proceeding with the EKC analysis, this issue has to be handled. Once the issue of data standardization is scrutinized, we will look into the issues of

---

[1]See Moomaw and Unruh (1997), Roberts and Grimes (1997), Roca and Alcántara (2001), Galeotti et al. (2006), Sinha (2015, 2016), Sinha and Bhattacharya (2016, 2017), Alam, Murad, Noman, and Ozturk (2016), Shahbaz, Lean, and Shabbir (2012, 2013a,b, 2016a,b,c).

Environmental Kuznets Curve (EKC). https://doi.org/10.1016/B978-0-12-816797-7.00007-2

variable selection and creation. First, while building an EKC model, it is necessary to choose the right variables. In doing so, the variables should confirm with the context, and thereafter, suitable transformations should be applied on the variables so that they can suit the model. Second, after choosing the right variables, it is necessary to optimize the number of variables, so that the problem of redundancy can be avoided. Third, given a particular context, it might be necessary to create variables, so that the impact of the parameters can be analyzed more effectively. Once all of these issues are addressed, then the issue of choosing the proxy for environmental quality or degradation comes to picture. In the literature of energy and environmental economics, carbon dioxide (CO_2) emission is the most used indicator of environmental degradation. However, there are several other environmental degradation indicators, which might also be chosen, and this choice should be driven by the geographic context chosen for the study.

By far, we have uncovered the issues, which are generally encountered by the researchers during the course of an EKC analysis. In this chapter, we have tried to address these issues with some basic and practical solutions. In doing so, we have analyzed some case studies and have demonstrated some examples of how to handle these issues. While addressing the issues using published studies as different cases, we have considered the data, which have actually been used by authors in those studies.

Rest of the chapter is designed in the following manner: Section 2 deals with choosing the proper EKC specification; Section 3 discusses the validation of an EKC model; Section 4 focuses on the issues of data standardization; Section 5 deals with the problem of variable selection and creation; Section 6 discusses choosing the proxy of environmental degradation; and Section 7 concludes the chapter.

CHOOSING THE PROPER ENVIRONMENTAL KUZNETS CURVE SPECIFICATION

In the literature of EKC analysis, predominantly two types of EKC models are visible, that is, quadratic and cubic specifications. Although EKC model can follow linear specification (Narayan & Narayan, 2010; Sinha & Shahbaz, 2018), criticism exists against this form of EKC. Brown and McDonough (2016) have stated that "The Narayan and Narayan (2010) approach may be thought of as similar to evaluating the time derivative of the elasticity of per capita CO_2 emissions with respect to per capita GDP with the idea of learning something about how CO_2 emissions vary with income or GDP. Unfortunately, such a time derivative indicates the relationship varies with time—not how it changes with income or GDP. Hence, evaluation of the time derivative does not provide any information about the shape of the EKC. (p. 456)" Because of this criticism, the linear model of EKC should be avoided for modeling purpose.

Now, we will look into the other two specifications of EKC, that is, quadratic and cubic. In case of quadratic specification of EKC, the form is generally an inverted U-shaped curve. During the preindustrial stage, the restricted utilization of natural resources and a circumscribed generation of recyclable wastes lower the extent of environmental degradation. With the rise in economic growth due to the rise in agriculture, rapid consumption of forest resources and the depletion of natural resources start exceeding their regeneration rates. Thereafter, acceleration of the industrialization process amplifies the magnitude and ferocity of the nonrecyclable waste. During the later stages of development process, increasing economic growth not only increases the returns but also wanes the environmental quality. Scale effect states that the technologic advancements provide increasing returns in terms of reduction in the pollution level (Torras & Boyce, 1998). In the due course of time, higher level of development causes growth in the knowledge intensive sectors and the sophistication of production technologies; these two circumstances have designated the composition and the technique effects, disparately (Fig. 7.1). The composition effect has a positive impact on environment, by transforming agrarian economy into a knowledge-intensive economy. This sectoral evolution, triggered by the composition effect, diminishes demand for energy consumption, and thereby, lessening the pollution level and drawing back the gradient of EKC (Halkos, 2003; Hettige, Mani, & Wheeler, 2000). Lastly, the technical effect implies the progresses in the production processes and the utilization of cleaner technologies, causing improvement in environmental quality level (Andreoni & Levinson, 2001). Besides, the occurrence of knowledge intensive sectors and the emergent ecologic threats may initiate stricter environmental laws, institutionalization of the green technologies, the substitution of superseded and contaminating technologies, and the supplementary environmental investments. This duration is followed by a steady and gradual decrease in the environmental degradation. Technical innovation is the key catalyst for this situation (Andreoni & Levinson, 2001). Therefore, it can be inferred that when technical

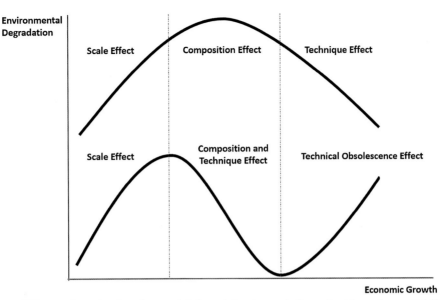

FIG. 7.1 Different channels of environmental degradation in quadratic and cubic environmental Kuznets curves.

innovation reinforces the endogenous aspect of the EKC hypothesis (Gradus & Smulders, 1993; van den Bergh & Nijkamp, 1994), the technical effect governs the income-emissions relationship by refining the production technologies and substituting the outmoded and contaminating technologies with cleaner technologies (Bovenberg & Smulders, 1995; Verdier, 1993).

Now, we will look into the cubic specification of EKC, that is, an N-shaped EKC. An N-shaped EKC pattern designates the association between income per capita and the level of environmental pollution in accordance with the EKC (Fig. 7.1). Even if pollution level falls with further economic growth, once the scale effect outstrips the technical effect, technical obsolescence ensues, along with the rises in the level of ambient pollution (Álvarez-Herranz, Balsalobre-Lorente, Cantos, & Shahbaz, 2017a, Álvarez-Herranz, Balsalobre-Lorente, Shahbaz, & Cantos, 2017b; Balsalobre & Alvarez, 2016). During the initial phase of economic growth, the prevailing development stage is recognized by the policies that generate biases, promote energy consumption, and bring about market failures (De Bruyn, van den Bergh, & Opschoor, 1998). In the subsequent phase, biases wane and market failures are adjusted. It is followed by the solicitation of authoritarian environmental policies and the rise in ecologic perception, which can be clarified by the developmental patterns pronounced by the EKC hypothesis

(Jones & Manuelli, 1995). Lastly, the relationship turns out to be N-shaped, when technical improvements do not have further growing yields, and the falling technical yields catalyze economies to return to a state of rising ecologic destruction (Balsalobre & Alvarez, 2016).

By far, we have discussed about the theoretic underpinning of the both forms of EKC. Now, the selection of these two specifications depends largely on contextual aspects, as these aspects demonstrate whether the technical obsolescence has set in, or not. While choosing a context, it should be checked that, whether there are any previous EKC studies that has been carried out on that context, or not. If the study has been carried out, then it is better to analyze the N-shaped EKC, as this form of EKC can show the nature of environmental degradation with further rise in income, which might not be possible with the inverted U-shaped EKCs. In case of no existing studies in that context, it is better to put forth a comparative scenario by choosing both the inverted U-shaped and the N-shaped EKC specifications. When the inverted U-shaped form can provide the present and an immediate futuristic aspect, the N-shaped form can provide a more futuristic aspect regarding the environmental degradation in that particular context. The results of both models can provide more flexibility to policy-makers during their short-term and long-term policy decisions.

VALIDATION PROBLEM OF AN ENVIRONMENTAL KUZNETS CURVE MODEL

The mathematical models of EKC should comply with certain conditions, which determine the shape and turnaround points, and thereby, confirm the empirical model. Through analysis of the literature, we discovered many studies that commented on the shape and turnaround points of the EKC without validation of the mathematical model derived by them. In this section, we focus on the cubic specification of EKC, that is, N-shaped and inverted N-shaped EKCs. The reason for choosing these two specifications is that the first-order conditions of these two specifications are in the quadratic form, and therefore it will be possible for us to show the validation criteria for these EKCs. For the inverted U-shaped EKCs, these validation criteria do not hold as the first-order condition in this case will be linear, and the maxima or minima value will be independent of income or the explanatory variable for economic growth. In this case, the validation of the model can be performed by looking at the signs of the coefficients of squared income or any other proxy of economic growth.[2] Moreover, the critical point of a quadratic specification, which is generally referred to as the inflection point, coincides with the turnaround point of the function. Therefore, the change in concavity of the model can be seen by looking at the turnaround point of a U-shaped or an inverted U-shaped EKC. This is not the case with an EKC with the cubic specification. For an EKC with the cubic specification, inflection point must lie between the two real turnaround points, as it reflects the change in concavity of the model. At level, a cubic mathematical model can result in the quadratic

first-order condition, which signifies the possibility of getting two real roots of the equation, as two turnaround points, whereas the second-order condition gives the inflection point. This happens for both N-shaped and inverted N-shaped specifications of EKCs. In both cases, the turnaround points must be real, and they should be derived from the first-order condition for the EKC to be valid. Therefore, the first differentiation of an EKC model must fulfill the basic validity criterion; however, this aspect has been ignored by some researchers while estimating the EKCs.

In this study, we have demonstrated how nonfulfillment of the validity condition for an EKC might lead to the falsification of that particular EKC. We have identified a number of studies that failed to consider this criterion. For example, Pal and Mitra (2017) estimated the EKCs for CO_2 emissions in India and China over the period 1971–2012. They found the shape of EKCs to be N-shaped only by looking at the sign of the coefficients of production: "The outcome shown in Table 7.4 indicates that for both countries the coefficients associated with the level values of production (Y_t) was positive, squared values of production (Y_t^2) was negative and the cubic values of production (Y_t^3) was positive. This supports the presence of an N–shaped EKC hypothesis …" (Pal & Mitra, 2017). It is evident from the example that the authors commented on the shape of the EKC only by looking at the signs of the coefficients of production and without testing the first- and second-order conditions of the mathematical model used. This is why it is essential to introduce the necessary condition for an N-shaped or an inverted N-shaped EKC to be valid. In doing so, first, we have noted the results derived by the authors, validated the models by using the data used in those studies, and then falsified those studies based on the validation criterion.

Let us assume that P is the indicator of pollution and Y is the indicator of economic growth. Then, a standard cubic specification of EKC takes the mathematical form in Eq. (7.1)

$$P = a_0 + a_1Y + a_2Y^2 + a_3Y^3 + \varepsilon \qquad (7.1)$$

From Eq. (7.1), we obtain the following specifications, which denote specific functional forms:
a. $a_1 = a_2 = a_3 = 0$; no growth-pollution association
b. $a_1 > 0$, $a_2 = a_3 = 0$; linearly increasing growth-pollution association
c. $a_1 < 0$, $a_2 = a_3 = 0$; linearly decreasing growth-pollution association
d. $a_1 > 0$, $a_2 < 0$, $a_3 = 0$; inverted U-shaped growth-pollution association
e. $a_1 < 0$, $a_2 > 0$, $a_3 = 0$; U-shaped/monotonically increasing growth-pollution association

---

[2]Let us assume that X is the indicator of pollution and that G is the indicator of economic growth. Then, a standard U-shaped/inverted U-shaped EKC takes the following mathematical form.

$$X = b_0 + b_1G + b_2G^2 + \varepsilon;$$

Or

$$\frac{dX}{dG} = b_1 + 2b_2G$$

Or,

$$\frac{d^2X}{dG^2} = 2b_2$$

$b_2 < 0$ implies the presence of local maxima, thereby indicating the evidence of inverted U-shaped EKC.
$b_2 > 0$ implies the presence of local minima, thereby indicating the evidence of U-shaped EKC.

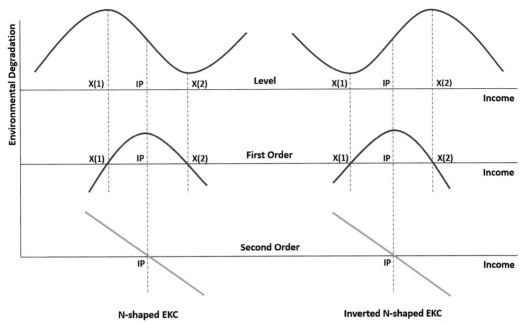

FIG. 7.2 Cubic specification of environmental Kuznets curve and corresponding turnaround and inflection points.

f. $a_1 > 0$, $a_2 < 0$, $a_3 > 0$; N-shaped growth-pollution association

g. $a_1 < 0$, $a_2 > 0$, $a_3 < 0$; inverted N-shaped growth-pollution association

The necessary condition for the EKC to be N-shaped is that $a_1, a_3 > 0$ and $a_2 < 0$. Similarly, for the EKC to be inverted N-shaped, the necessary condition is $a_1, a_3 < 0$ and $a_2 > 0$. However, this condition is not sufficient for deciding on the nature of the EKC, as this condition does not reflect anything about the validity of the model. To check the validity of the model, the model should be differentiated to the first order. The first-order differential of Eq. (7.1) is given by

$$\frac{dP}{dY} = a_1 + 2a_2 Y + 3a_3 Y^2 = 0 \qquad (7.2)$$

For the EKC to be N-shaped or inverted N-shaped, Eq. (7.1) must have local maxima and minima at two distinct values of Y.[3] The condition for Eq. (7.1) having local maxima and minima is given by Eq. (7.2):

$$a_2^2 - 3a_1 a_3 > 0 \qquad (7.3)$$

[3]The local maxima and minima can be found at $Y = \left(-2a_2 \pm \sqrt{4a_2^2 - 12a_1 a_3} \right) / 6a_3$, or, $Y = \left(-a_2 \pm \sqrt{a_2^2 - 3a_1 a_3} \right) / 3a_3$. These are derived by solving the first order condition given in Eq. (7.2).

To find the values of the maxima and minima, arriving at the second-order condition is required. The second-order condition, derived from Eq. (7.2), takes the following form:

$$\frac{d^2 P}{dY^2} = 2a_2 + 6a_3 Y = \pm \sqrt{4a_2^2 - 12a_1 a_3} \qquad (7.4)$$

The validity of the second-order condition is also provided by Eq. (7.3). Therefore, it can be stated that Eq. (7.3) is the sufficient condition for an N-shaped or an inverted N-shaped EKC to be valid.[4] Therefore, for an N-shaped EKC, the two conditions are (1) a_1, $a_3 > 0$, $a_2 < 0$, and (2) $a_2^2 - 3 a_1 a_3 > 0$. Similarly, for an inverted N-shaped EKC, the two conditions are (1) $a_1, a_3 < 0$, $a_2 > 0$ and (2) $a_2^2 - 3 a_1 a_3 > 0$. If it is found that for any given model, the first condition holds, but the second condition does not hold, then the EKC can never be estimated, as the turnaround points will not be real. However, in the existing literature of energy and environmental economics, there are several instances where researchers commented on the shape of the EKC by looking at the first condition only.

[4]$Y_{maxima} > Y_{minima}$: EKC is N-shaped.
$Y_{maxima} < Y_{minima}$: EKC is inverted N-shaped.
$Y_{maxima} = Y_{minima}$: EKC is U shaped/inverted U-shaped.

To bring forth more clarity in the aforementioned discussion, let us look into the physical interpretation of the inflection point and the turnaround points. In Fig. 7.2, the cubic specifications of the EKC, along with their corresponding turnaround points and the inflection point have been demonstrated. For both the N-shaped and the inverted N-shaped EKCs, the first-order condition is of quadratic specification, as it has already been shown in Eq. (7.2). Now, to EKCs to have the real turnaround points, the first-order condition must have two real roots. Let as assume that the roots of the first-order equation are $X(1)$ and $X(2)$. As these values are the real turnaround points, then $Y \in [X(1), X(2)]$. After putting these values in Eq. (7.1), if $P_{X(1)} > P_{X(2)}$, then the EKC is an N-shaped, and if $P_{X(1)} < P_{X(2)}$, then the EKC is an inverted N-shaped. Now, equating the second-order condition, that is, Eq. (7.4) to zero gives the inflection point, as per the following:

$$\frac{d^2P}{dY^2} = 2a_2 + 6a_3Y = 0 \qquad (7.5)$$

or

$$Y = -a_2/3a_3 \qquad (7.6)$$

The value of Y received in Eq. (7.6) is the value of the inflection point for an EKC with cubic specification. This value is the evidence of the real turnaround points, as this point can exist only between two real turnaround points. This point is where the straight line in Eq. (7.5) intersects the horizontal axis. Therefore, it can be stated that to an N-shaped or an inverted N-shaped EKC to be valid, the inflection point must exist, which can only be fulfilled, if the straight line in Eq. (7.5) intersects the horizontal axis. This geometric intersection actually takes us back to Eq. (7.3) and Eq. (7.4), which we derived as the second-order condition of the EKC model for achieving the maxima and minima. The conditions for achieving the real turnaround points actually indicates the existence of the inflection point, as without it, the concavity of an N-shaped or an inverted N-shaped EKC can never change. Therefore, if any EKC has $(a_1, a_3 > 0, a_2 < 0)$ or $(a_1, a_3 < 0, a_2 > 0)$, then apparently, it will be erroneous to comment on the shape of the EKC, as this condition does not say anything about the second condition, which is necessary for the evidence of the inflection point.

We have by far discussed how two conditions need to be fulfilled before commenting on the shape of an EKC with cubic specification. In doing so, we have identified a number of studies where the researchers stated the models to be valid, but they failed to fulfill the second condition. We have tested the models derived by the authors and found that $a_2^2 - 3 a_1 a_3$ comes out to be less than zero, as indicated in Table 7.1. Nonfulfillment of these conditions will lead to the turnaround points, which lie on the imaginary plane rather than Cartesian plane, as identified by some researchers (Abid, 2016; Sebri, 2016).

PROBLEM OF DATA IN ENVIRONMENTAL KUZNETS CURVE MODEL

By far, EKC studies have largely used three types of data, namely cross-sectional, time series, and panel. The first-generation studies (e.g., Panayotou, 1993) have majorly used the cross-sectional data for EKC analysis. However, in the course of the time, the usage of time series and panel data started gaining prominence among the researchers. For example, between 1991 and 2017, considering the EKC estimation for CO_2 emissions, nearly 144 studies have been published with the time-series data, and 124 studies have been published with the panel data (Shahbaz & Sinha, 2019). This shows the popularity of these two types of data among the researchers for the analysis of the EKC.

While talking about the data requirements for an EKC model, it should be remembered that the EKC analysis is carried out to look for the impact of economic growth on the environmental quality over different phases of economic development. In such a scenario, using cross-sectional data for EKC analysis might not solve the purpose, as the EKC analysis requires environmental quality to be a flow variable, not a stock variable. Thus, using time series or panel data is more suitable in the EKC modeling. During the analysis of the EKC for a single context, time-series data are used, whereas panel data are used for cross-contextual (country) analysis. The latter can provide us with a comparative analysis, whereas the former can provide us with an analysis carried out in a single geographic context.

Now, if the data used in the EKC analysis are analyzed, it could be seen that the data suffer from a number of problems, and these problems need to be addressed before carrying out the analysis. We will discuss the problems in the following subsections.

Conversion of Base Year

The data used in the EKC modeling might not be standardized in terms of the base year. Before carrying out the empirical analysis, all the variables should be brought to one common base year, as it will ensure the nearly similar temporal impact on all the variables, and the EKC can be analyzed for a given period of time. For example, gross state domestic product (GSDP) data for Indian states are available for different base years,

TABLE 7.1
Evidence of the Falsified Environmental Kuznets Curve (EKC) Studies.

Author(s)	Model Derived by Author(s)	Reported Verdict	$a_2^2 - 3a_1a_3$	Our Verdict
Martínez-Zarzoso and Bengochea-Morancho (2004)	$LC = 94.1*LY - 12.43*LY^2 + 0.56*LY^3$	N-shaped EKC	-3.5831	Invalid model
Dijkgraaf and Vollebergh (2005)	$C = -142.2*Y + 5.05*Y^2 - 0.19*Y^3$	Inverted U-shaped EKC	-55.5515	Invalid model
Galeotti and Lanza (2005)	$LC = 4.435*LY - 0.394*LY^2 + 0.012*LY^3$	N-shaped EKC	-0.0044	Invalid model
Galeotti et al. (2006)	$C = 7.2*Y - 0.74*Y^2 + 0.03*Y^3$	Model cannot be rejected	-0.1004	Invalid model
	$C = 17.62*Y - 2.14*Y^2 + 0.09*Y^3$	Model cannot be rejected	-0.1778	Invalid model
Bagliani, Bravo, and Dalmazzone (2008)	$EF = 0.0006*Y - 2.26E-08*Y^2 + 2.94E-13*Y^3$	N-shaped EKC	-1.844E-17	Invalid model
	$EF = 0.0004*Y - 1.34E-08*Y^2 + 1.57E-13*Y^3$	N-shaped EKC	-8.84E-18	Invalid model
	$EF = 0.0002*Y - 8.51E-09*Y^2 + 1.25E-13*Y^3$	N-shaped EKC	-2.58E-18	Invalid model
Brajer, Mead, and Xiao (2008)	$LS = 8.78*LY - 0.899*LY^2 + 0.031*LY^3$	N-shaped EKC	-0.0083	Invalid model
Mazzanti, Montini, and Zoboli (2008)	$PM_{10} = 4.675*VA - 1.008* VA^2 + 0.074* VA^3$	N-shaped EKC	-0.0218	Invalid model
Akbostanci et al. (2009)	$C = 3.5597*Y - 2.3475*Y^2 + 0.54145*Y^3$	N-shaped EKC	-0.2714	Invalid model
	$LC = 3.3647*LY - 6.2111*LY^2 + 3.82435*LY^3$	N-shaped EKC	-0.0256	Invalid model
Lee, Chiu, and Sun (2009)	$C = 0.324*Y - 0.012*Y^2 + 0.00015*Y^3$	N-shaped EKC	-1.8E-06	Invalid model
	$C = 0.349*Y - 0.013*Y^2 + 0.000172*Y^3$	N-shaped EKC	-1.108E-05	Invalid model
	$C = 0.341*Y - 0.013*Y^2 + 0.000166*Y^3$	N-shaped EKC	-8.18E-07	Invalid model
Mohapatra and Giri (2009)	$N = 0.0031*Y - 3.963E-07*Y^2 + 1.853E-11*Y^3$	N-shaped EKC	-1.961E-14	Invalid model
	$N = 0.0156*Y - 9.193E-07*Y^2 + 1.815E-11 *Y^3$	N-shaped EKC	-4.307E-15	Invalid model
He and Richard (2010)	$C = 3.4950*Y - 0.1060*Y^2 + 0.0012*Y^3$	N-shaped EKC	-0.0013	Invalid model
	$C = 3.1698*Y - 0.1011*Y^2 + 0.0011*Y^3$	N-shaped EKC	-0.0002	Invalid model
Lipford and Yandle (2010)	$C = 131.99*Y - 0.0054*Y^2 + 0.0000000818*Y^3$	N-shaped EKC	-0.00000323	Invalid model
	$C = 75.1*Y - 0.003*Y^2 + 0.0000000483*Y^3$	N-shaped EKC	-0.00000188	Invalid model
	$C = 138.22*Y - 0.0058*Y^2 + 0.00000011*Y^3$	N-shaped EKC	-0.00001197	Invalid model
	$C = 1159.12*Y - 0.00417*Y^2 + 0.000000525*Y^3$	N-shaped EKC	-0.00180823	Invalid model
	$C = 3452.24*Y - 1.0887*Y^2 + 0.000121*Y^3$	N-shaped EKC	-0.06789543	Invalid model
	$C = 0.00146*Y - 0.000000064*Y^2 + 1.06E-12*Y^3$	N-shaped EKC	-5.468E-16	Invalid model
	$C = 0.00149*Y - 0.000000758*Y^2 + 1.45E-12*Y^3$	N-shaped EKC	-7.358E-16	Invalid model
	$C = 0.00295*Y - 0.000000972*Y^2 + 1.08E-10*Y^3$	N-shaped EKC	-1.101E-14	Invalid model

Continued

TABLE 7.1
Evidence of the Falsified Environmental Kuznets Curve (EKC) Studies.—cont'd

Author(s)	Model Derived by Author(s)	Reported Verdict	$a_2^2 - 3a_1a_3$	Our Verdict
Brajer, Mead, and Xiao (2011)	$LS = 8.796*LY - 0.946*LY^2 + 0.034*LY^3$	N-shaped EKC	−0.0022	Invalid model
Sinha Babu and Datta (2013)	$EDI = 12.919*DBI - 18.52538* DBI^2 + 8.975398* DBI^3$	N-shaped EKC	−4.6698	Invalid model
Onafowora and Owoye (2014)	$LC = 5.034*LY - 0.813*LY^2 + 1.042*LY^3$	N-shaped EKC	−15.0753	Invalid model
	$LC = 3.489*LY - 0.615*LY^2 + 0.605*LY^3$	N-shaped EKC	−5.9543	Invalid model
	$LC = 9.605*LY - 1.710*LY^2 + 0.630*LY^3$	N-shaped EKC	−15.2294	Invalid model
	$LC = 3.275*LY - 0.535*LY^2 + 0.566*LY^3$	N-shaped EKC	−5.2747	Invalid model
	$LC = 4.728*LY - 0.677*LY^2 + 0.405*LY^3$	N-shaped EKC	−5.2862	Invalid model
	$LC = 4.895*LY - 0.781*LY^2 + 0.659*LY^3$	N-shaped EKC	−9.0675	Invalid model
Balibey (2015)	$LC = 240.4722*LY - 27.17071*LY^2 + 1.027006*LY^3$	N-shaped EKC	−2.6517	Invalid model
Dong, Wang, and Guo (2016)	$C = 0.142*Y - 2.36E{-}6*Y^2 + 1.36E{-}11*Y^3$	N-shaped EKC	−2.24E−13	Invalid model
Uddin, Alam, and Gow (2016)	$EF = 0.32*Y - 1.23*Y^2 + 9.81*Y^3$	N-shaped EKC	−7.9047	Invalid model
	$EF = -1.53*Y + 2.98*Y^2 - 4.92*Y^3$	Inverted N-shaped EKC	−13.7024	Invalid model
	$EF = 3.49*Y - 3.23*Y^2 + 4.96*Y^3$	N-shaped EKC	−41.4983	Invalid model
	$EF = 13.96*Y - 2.47*Y^2 + 5.16*Y^3$	N-shaped EKC	−209.9999	Invalid model
	$EF = -5.06*Y + 1.13*Y^2 - 0.97*Y^3$	Inverted N-shaped EKC	−13.4477	Invalid model
	$EF = -1.706*Y + 0.049*Y^2 - 0.046*Y^3$	N-shaped EKC	−0.2330	Invalid model
Abid (2017)	$C = 11.142*Y - 1.186*Y^2 + 0.043*Y^3$	N-shaped EKC	−0.0307	Invalid model
Boamah et al. (2017)	$C = 3.00*Y - 0.41*Y^2 + 0.02*Y^3$	N-shaped EKC	−0.0119	Invalid model
das Neves Almeida, Cruz, Barata, and García-Sánchez (2017)	$CIEP = 5.9825739*Y - 9.352259*Y^2 + 5.0266098*Y^3$	N-shaped EKC	−2.7514	Invalid model
Pal and Mitra (2017)	$LC = 7.8345*LY - 2.8164*LY^2 + 3.3789*LY^3$	N-shaped EKC	−71.4843	Invalid model

Note: C, CO_2 emissions; CIEP, Composite Index of Environmental Performance; DBI, Development Balance Index.; EDI, Environmental Degradation Index; EF, ecologic footprint; LC, log(CO_2 emissions); LN, log(NO_2 emissions); LS, log(SO_2 emissions); N, NO_2 emissions; PM_{10}, particulate matter; S, SO_2 emissions; VA, value added.

that is, 1999–2000, 2004–05, and so on. Now, if we want to conduct any study for the period of 2001–13 for Indian states, the base years have to be changed to the latest base year. This method has been followed in Sinha (2016), Sinha and Bhattacharya (2016, 2017), and Sinha and Rastogi (2017) for different base years. Let us assume that 2004–05 is the latest base year, given the time period 2001–13, and we convert the data by changing the base year from 1999–2000 to 2004–05. A working example of this particular conversion of base year has been shown in Table 7.2. Here, we have taken the GSDP data for one Indian state (Andhra Pradesh is taken as an example) to demonstrate the exercise.

We will now explain the working of the method. First, we will have to choose the latest base year, according to which the data will be transformed. In this example, 2004–05 is the base year, and these data will be used as the reference for this exercise. Second, we will have to transform the data points of the previous base year, that is, 1999–2000 in this case, for which data are not available in the current base year. In this example, such data points are between 2001 and 2003, and these three data points will only be converted according to the new base year. Third, for the conversion, we will have to select the data points for the years, in which the data are available for both the base years. Then we will take the ratio of the data points of the latest and old base years, and those ratios will be averaged.

This average value is the conversion factor for the data points we have mentioned in the second step. Fourth, the conversion factor hence obtained will be multiplied with the data points in the base year 1999–2000 for the period of 2001–03. These data points are now the converted data points for the base year 2004–05. Finally, the last column of Table 7.2 contains the final data points to be used, and in these data points, first three data points are the converted ones, and the remaining data points are the original data points for the base year 2004–05.

Handling Missing Data Points

While carrying out an EKC estimation exercise, it has been seen that many variables do not have continuous data for all the contexts. Now, there are a number of studies in the literature, which have used unbalanced panel data (e.g., Allard, Takman, Uddin, & Ahmed, 2018; Caviglia-Harris, Chambers, & Kahn, 2009; Ercolano, Gaeta, Ghinoi, & Silvestri, 2018; Pao & Tsai, 2010, 2011; Wang, Zhang, Kubota, Zhu, & Lu, 2015). However, in most of these studies, we do not find any evidence on how the unbalanced nature of the panel data has been addressed. Moreover, the researchers have not focused on the problems they have faced. Baltagi (2005) have identified some of the problems regarding unbalanced panel data. The major problem that unbalanced panel data can cause is the bringing

TABLE 7.2
Conversion of Base Year Exercise.

Year	Base Year 1999–2000 (A)	Base Year 2004–2005 (B)	(C) = (B)/(A)	(D) = (A) * Average of (C)	Final Data
2001	145,185	—	—	185,917	185,917
2002	149,142	—	—	190,984	190,984
2003	163,085	—	—	208,838	208,838
2004	176,379	224,713	1.27403	—	224,713
2005	194,437	246,210	1.26627	—	246,210
2006	216,140	273,730	1.26645	—	273,730
2007	239,372	306,645	1.28104	—	306,645
2008	251,431	327,731	1.30346	—	327,731
2009	265,140	342,571	1.29204	—	342,571
2010	—	382,459	—	—	382,459
2011	—	411,184	—	—	411,184
2012	—	432,112	—	—	432,112
2013	—	455,937	—	—	455,937

Directorate of Economic and Statistics of Andhra Pradesh, India.

TABLE 7.3
Growth Rate Method of Interpolation.

Year	Census Data	Growth Rate	Interpolated Data		New Series
2001	3,735,202	(Census data$_{2011}$ − Census data$_{2001}$)/10 = (4174064 − 3735202)/10 ≈ 43886	3,735,202	−	3,735,202
2002	−		3735202 + 43886 =	3,779,088	3,779,088
2003	−		3779088 + 43886 =	3,822,974	3,822,974
2004	−		3822974 + 43886 =	3,866,860	3,866,860
2005	−		3866860 + 43886 =	3,910,746	3,910,746
2006	−		3910746 + 43886 =	3,954,632	3,954,632
2007	−		3954632 + 43886 =	3,998,518	3,998,518
2008	−		3998518 + 43886 =	4,042,404	4,042,404
2009	−		4042404 + 43886 =	4,086,290	4,086,290
2010	−		4086290 + 43886 =	4,130,176	4,130,176
2011	4,174,064		4,174,064	−	4,174,064
2012	−		4174064 + 43886 =	4,217,948	4,217,948
2013	−		4217948 + 43886 =	4,261,834	4,261,834

Census, India.

more randomness in the stochastic error term, and thereby, inflating the error term.

One of the solutions to this problem is to remove the specific period(s) from the study, for which the data are missing. However, this will reduce the dimension of the panel, and thereafter, it will reduce the explanatory power of the data. This might have an impact on the results of analysis, as the degree of freedom will change. Therefore, it is not advisable to carry out this procedure. Now, Biørn (2004) suggested a number of procedures to be followed for carrying out analysis on an unbalanced panel data set. The prerequisite of these methods is that the missing data points should comply with the assumption of Missing Completely at Random (MCAR). When we look at the country-level data for carrying out an EKC analysis and we come across missing data points, then we can see that they are not randomly missing, rather the stochastic trend in the missing data points can be observed. In such a scenario, it is advisable to fill up the data points by linear interpolation and extrapolation, or sometimes by simple or moving average.

One of the classic examples in this case is the missing data points for Indian population data, which is collected in every 10 years. Therefore, the population data for Indian states are always having data gaps for a decade. Moreover, the application of linear interpolation approach, the method used to fill up the data gaps in this case is the "growth rate" approach. In this method, we compute the annual growth rate for the decade and fill up the series based on that computed growth rate by means of interpolation. This method has been followed in Sinha (2016), Sinha and

Bhattacharya (2016, 2017), and Sinha and Rastogi (2017). An illustrious example is provided in Table 7.3. Let us assume that our study period is 2001–13. We have the population data from Indian Census for 2001 and 2011, and Chittoor[5] has been taken as an example. Using these two data points, we will build the entire data set for Chittoor over the period of 2001–13. Let us demonstrate the method with this example in Table 7.3.

The example demonstrates how missing data can be filled up using the "growth rate" approach. Now, it can be argued that how the annual growth computed between 2001 and 2011 can be applied for the series post-2011. The reason behind this application is that the population data for the decade 2011–21 are not available yet, and therefore it is an assumption that the population growth rate remains constant across the two decades.

PROBLEM OF VARIABLE SELECTION IN ENVIRONMENTAL KUZNETS CURVE MODEL

To bring forth more insights and robustness in the study, researchers have an affinity to include more variables in the EKC model. Nature of these variables changes along with the changes in the global and local contexts. For example, international trade has been an integral part of the EKC modeling since the seminal work of Grossman and Krueger (1991). Later on, the aspects of energy consumption started being an important part of the EKC models. The study by Cole, Rayner, and

---

[5]A city in the district of Andhra Pradesh, India.

Bates (1997) was one of the earliest studies to incorporate fossil fuel energy consumption in the EKC modeling exercise. With the graduation of time, renewable energy consumption started to gain prominence in the global ecologic scenario, and it took almost a decade to incorporate this variable in the EKC models. The study by Richmond and Kaufmann (2006) was one of the earliest studies to incorporate renewable energy consumption in the EKC modeling approach. Apart from trade and energy consumption, financial development of a nation is also important to determine the national level of environmental degradation. Although Dasgupta, Laplante, and Mamingi (2001) have identified this issue, Tamazian, Chousa, and Vadlamannati (2009) introduced this aspect in the EKC models.

Following the course of this discussion, it can be inferred that there are several variables, which have been included in the EKC literature. There exist some problems with the selection and the creation of variables. First, using the right variables is crucial during the analysis of EKC. Second, in any EKC estimation exercise, optimum number of variables needs to be chosen. Third, the researchers need to create necessary variables to check the robustness of the results in the EKC analysis. We will discuss these problems in the following subsections.

Problem of Selecting Right Variables

Over the last decade, EKC analysis has been carried out on a number of contexts, which are both single country and cross-country. Along with the geographic contexts (single or cross-country), the temporal contexts (study period) exist. Therefore, the chosen variables must comply with these two context settings. An elementary EKC model consists of the indicators of economic growth, that is, per capital income or GDP, and the indicator of environmental degradation, that is, various pollutants, water and land contamination, ecologic footprint, and various others. Now, to bring forth width and depth in the analysis, and to address the omitted variable bias, more variables are needed to be added in the EKC model (Stern, 2004). This addition of variables should not be carried out at random, as the EKC model might suffer from misspecification bias. To avoid this issue, the context should be analyzed thoroughly, and then the variables can be chosen for the analysis.

We will now demonstrate this aspect with some examples. Let us begin with the studies carried out by Sinha and Bhattacharya (2016, 2017). In this study, apart from income, other explanatory variables are petroleum consumption, electricity consumption, and population. The studies were carried out for 139 Indian cities over the period 2001−13. Here, the explanatory variables were chosen in keeping with the context setting. Sinha and Bhattacharya (2016) have mentioned, "For Indian cities, most of the human made ambient air pollution can be attributed to fossil fuel consumption, and it can be further subdivided into commercial electricity consumption and combustion of petroleum products. Majority of the commercial electricity utilized in India is generated out of the thermal power plants, which run on coal and crude oil. On the other hand, direct utilization of petroleum products can be seen in transportation sector and in households (p. 3)." It states that how the explanatory variables should comply with the context setting.

Now, we will look into the study by Shahbaz, Nasir, and Roubaud (2018b). This study was carried out for France over the period of 1955−2016. Alongside income, foreign direct investment (FDI), and energy consumption, energy innovation was used to indicate real research and development expenditures for energy innovations. While choosing this variable, it has been mentioned in the study that the inflow of technologic innovations by means of trade can improve the environmental quality, and for the case of France, this was observable. Moreover, the public and private corporations in France have been increasingly investing in the research and development processes in the pursuit of cleaner technologies. In view of this evidence, it becomes clear the intention behind the choice of this variable, which was driven by the context setting.

Saying this, during selection of the variables for an EKC model in any given context, it is necessary to analyze the particular contextual aspects, which might affect environmental quality in that context. For example, if we consider the context of any developing or emerging economy, then urbanization should be one of the explanatory variables in the EKC model. In some cases, corruption index may be an important variable in the similar context, as the incidents of corruptive practices regarding the protection of right for public goods are prevalent in these countries. Moreover, the incidental impact of any major global crisis can be captured through dummy variables. The last one is applicable for the cases of the developed nations, as well.

From the methodologic perspective, it should be remembered that the explanatory variables should not be correlated; otherwise, the results might be spurious. For example, choosing per capita income, population, and per capita energy consumption might create problem, as these variables are highly correlated. Saying this, it should also be noted that removal of the correlation might make these variables perfectly applicable for the analysis. Sinha and Bhattacharya (2016, 2017) have applied the orthogonal transformation on the

TABLE 7.4
Correlation Matrix of Original and Orthogonally Transformed Variables.

	Original Variables				Transformed Variables		
	Y	P	E		Y	P	E
Y	1.0000			Y	1.0000		
P	0.9503	1.0000		P	0.0000	1.0000	
E	0.9613	0.8920	1.0000	E	−0.0000	0.0000	1.0000

Y, per capita income; *P*, population; *E*, energy consumption.

correlated variables to get rid of this issue. In Table 7.4, we will demonstrate that exercise.

In the exercise demonstrated in Table 7.4, it can be shown that the transformed variables are perfectly fit to be included in the analysis, as there are no correlations among them. This is one of the ways to include the correlation-corrected variables in an EKC model.

Problem of Choosing the Number of Variables

In the recent EKC literature, it has been seen that the EKC models generally use a number of variables to bring forth depth in the analysis. Therefore, it is also necessary to discuss that there should be a certain number of variables, beyond which additional variables cannot be added in an EKC model. As EKC modeling approach is different than the statistic prediction, adding more predictors might reduce the statistical significance of the individual predictors.

Although there are several techniques in the literature to determine the optimum variables to be included in a model, we will demonstrate the Least Angle Regression with Lasso Algorithm devised by Efron, Hastie, Johnstone, and Tibshirani (2004). This method can be used as variable reduction method. One of its major advantages is that it can offer the property of parsimony by suppressing the insignificant coefficients to zero, and thereby, removing the variables.

This will be clear, if we demonstrate the method with an example. Let us assume that we want to estimate the following EKC model for NO$_2$ emissions in 139 Indian cities over the period of 2001−13:

$$P_{it} = C_i + \sum_{j=1}^{3} \alpha_j Y_{jt}^j + \sum_{k=1}^{3} \alpha_{k+3} Pop_{kt}^k + \alpha_7 Gen_{it} + \alpha_8 EC_{it}$$
$$+ \alpha_9 LR_{it} + \alpha_{10} CG_{it} + \alpha_{11} CG_{it} Y_{it} + \alpha_{12} CG_{it} Gen_{it}$$
$$+ \alpha_{13} CG_{it} EC_{it} + \alpha_{14} CG_{it} LR_{it} + \alpha_{15} CG_{it} Y_{it} Gen_{it}$$
$$+ \alpha_{16} CG_{it} Y_{it} EC_{it} + \alpha_{17} CG_{it} Y_{it} LR_{it} + \varepsilon_{it}$$
$$(7.7)$$

where for city *i* in year *t*, P_{it} stands for the level of emissions, Y_{it} is the level of income at city level, Pop_{it} is the population, CG_{it} is the indicator of political collaboration between state and central government, Gen_{it} stands for the gender ratio in terms of number of women per 1000 men, EC_{it} is the consumption of electricity, LR_{it} is the literacy rate, ε_{it} is the error term, and C_i is the city level fixed effect. As it can be seen that the EKC model in Eq. (7.7) has 17 explanatory variables, the optimum number of variables in the model has to be decided. To achieve this aim, we have applied the Least Angle Regression with Lasso Algorithm. In Table 7.5, the results of this method are tabulated.

As seen in Table 7.5, it can be seen that the prediction error is least at step 15, and after this step, the prediction error and the R-square in marginal start increasing. Therefore, the variables added in the model after step 15 will not be considered in the model, that is, CG.Y, CG.Gen, and CG.LR will not be considered in the model. Thus, the final number of explanatory variables to be considered in the model comes out to be 14.

Various researchers in diverse ecologic contexts (e.g., Papież, Śmiech, & Frodyma, 2018; Tan, Liu, Li, & Wang, 2018; Ver Hoef, Peterson, Hooten, Hanks, & Fortin, 2018) have used this method. However, this method has yet to be used in optimizing the number of variables in the EKC analyses. Therefore, this variable reduction method can be used extensively to optimize the EKC models.

Problem of Variable Creation

In the literature of the EKC analysis, there are several studies, which have introduced new variables, to bring forth additional insights to their studies. Now, these insights can be expected in terms of point elasticity of the environmental degradation with respect to the variables. Researchers have used dummy variables extensively for obtaining these elasticities. In this connection, it is important to mention that, usage of dummy variables has a long history in the economic

TABLE 7.5
Results of Least Angle Regression With Lasso Algorithm.

Step	Prediction Error	R-square	Action
1	534.7515	0.0000	—
2	296.5923	0.1527	Add Gen
3	264.4790	0.1744	Add Y^2
4	232.6340	0.1959	Add Pop^2
5	180.6471	0.2302	Add LR
6	179.7542	0.2321	Add EC
7	56.9395	0.3114	Add Y
8	45.7538	0.3198	Add Pop^3
9	39.0985	0.3253	Add CG.Y.LR
10	27.1250	0.3342	Add CG.EC
11	15.2305	0.3430	Add CG.Y.EC
12	17.1748	0.3431	Add CG.Y.Gen
13	18.5794	0.3435	Add Pop
14	18.9295	0.3445	Add Y^3
15	12.5112 *	0.3499	Add CG
16	14.2475	0.3500	Add CG.Y
17	16.2188	0.3500	Add CG.Gen
18	18.0000	0.3502	Add CG.LR

Central Pollution Control Board, India.

literature. However, for carrying out the EKC analysis, variable creation should not be restricted solely to the dummy variables. Henceforth, the interaction variables turn out to be necessary. Calculation of elasticity calls for the first differentiation of the model variables, and while going for differentiation, the intended variables remain intact because of this interaction effect.

This aspect will be clear, when the entire discussion is put mathematically. Let us extend Eq. (7.1) by including an explanatory variable X and an interaction of X with Y.

$$P = a_0 + a_1 Y + a_2 Y^2 + a_3 Y^3 + a_4 X + a_5 XY + \varepsilon \quad (7.8)$$

$$\frac{\partial P}{\partial Y} = a_1 + 2a_2 Y + 3a_3 Y^2 + a_5 X \quad (7.9)$$

$$\left. \frac{\partial P}{\partial Y} \right|_{Y=0} = a_5 X \quad (7.10)$$

Now, Eq. (7.10) elucidates the point elasticity of the pollution with respect to income at $Y = 0$, and the real value of the point elasticity can be obtained because of

the interaction variable XY. This point elasticity can also be handled from the perspective of X. We will start with Eq. (7.8) and differentiate it with respect to X, and then check the point elasticity.

$$\frac{\partial P}{\partial X} = a_4 + a_5 Y \quad (7.11)$$

$$\left. \frac{\partial P}{\partial X} \right|_{X=0} = \frac{\partial P}{\partial X} = a_4 + a_5 Y \quad (7.12)$$

Now, Eq. (7.12) clarifies the point elasticity of the pollution with respect to X at $X = 0$, and the real value of the point elasticity can be obtained because of the interaction variable XY. In the absence of interaction variable, X is retained owing to the interaction variable. However, in the case of the logarithmic values of the variables, which is actually the case of any OLS-based EKC model, the treatment could have been different. Rewriting Eq. (7.8) in logarithmic form, the further proceedings will be as per the following:

$$lnP = a_0 + a_1 lnY + a_2 (lnY)^2 + a_3 (lnY)^3 + a_4 lnX \\ + a_5 lnX.lnY + \varepsilon \quad (7.13)$$

$$\frac{\partial P}{P} = a_1 \frac{\partial Y}{Y} + 2a_2 (lnY)\frac{\partial Y}{Y} + 3a_3 (lnY)^2 \frac{\partial Y}{Y} + a_4 \frac{\partial X}{X} \\ + a_5 \left(lnX \frac{\partial Y}{Y} + lnY \frac{\partial X}{X} \right) \quad (7.14)$$

Given this, the elasticity of the pollution with respect to X comes out to be as per the following:

$$\frac{\partial P}{P} \bigg/ \frac{\partial X}{X} = a_4 + a_5 lnY \quad (7.15)$$

This value is similar to the value obtained in Eq. (7.12). This is the value of elasticity to be obtained in any EKC model by using the interaction variable. Without the interaction variable, the elasticity value would be a_4. Therefore, we may have an idea about the additional information we can obtain from the usage of interaction variable.

Once we have discussed the mathematical interpretation of the interaction variables, we should also discuss about when this interaction variable can be used. This variable is used, when the impact of this variable can be seen directly on the other endogenous model variables, that is, income, in this case. For example, in Sinha, Shahbaz, and Balsalobre (2017), the interaction between income and biomass energy consumption was used, as it has been seen that utilization of better quality biomass energy is catalyzed by income level. Moreover, this study also used the interaction between income and trade openness, as it has been seen that the impact of technologic innovation, which was indicated by trade openness, goes up with the level of income. As the chosen context has shown the mentioned characteristics, the creation of interaction variables was theoretically justified.

In this context, the study by Shahbaz, Balsalobre, and Shahzad (2018a) needs a special mention. This study applied six interaction variables across 12 EKC models: (1) interaction between institutional quality and real foreign direct investment per capita, (2) interaction between institutional quality and real trade per capita, (3) interaction between institutional quality and economic globalization, (4) interaction between institutional quality and social globalization, (5) interaction between institutional quality and political globalization, and (6) interaction between institutional quality and overall globalization. Considering the context of G7 economies, these interaction variables were significant and relevant to the context, as the institutional quality plays an important role in shaping the environmental degradation through various international channels in G7 countries.

CHOOSING THE PROXY FOR ENVIRONMENTAL DEGRADATION

In the literature of EKC analysis, it has been observed that researchers are more prone toward considering CO_2 emissions as the main proxy for environmental degradation. There are several reasons behind this particular choice made by the researchers. First, the ease of availability of CO_2 emissions data helps researchers to take up the EKC analysis in diverse contexts. Second, having the largest contribution in the greenhouse gas (GHG) emissions, CO_2 can be considered as a convenient proxy for the environmental degradation. Third, because CO_2 is the most prominent global pollutant among the GHGs, it has a higher acceptability in the cross–country EKC studies (Miah, Masum, & Koike, 2010).

However, in the literature of EKC analysis, CO_2 is not the only proxy of environmental degradation used by the researchers. Apart from CO_2, there are other proxies, which have been used for environmental degradation. Studies have considered nitrogenous oxide (NO_X) emissions (Archibald, Banu, & Bochniarz, 2004; Carson, Jeon, & McCubbin, 1997; Egli, 2001; Fonkych & Lempert, 2005; Grossman & Krueger, 1995; Hill & Magnani, 2002; Panayotou, 1993; Park & Lee, 2011; Roumasset, Burnett, & Wang, 2006; Song, Zhang, & Wang, 2013; Welsch, 2004), sulfur dioxide (SO_2) emissions (Akbostancı, Türüt-Aşık, & Tunç, 2009; Apergis & Payne, 2009; Al

Sayed & Sek, 2013; Bimonte, 2002; Galeotti, Lanza, & Pauli, 2006; Grossman & Krueger, 1995; Harbaugh, Levinson, & Wilson, 2002; Kaufmann, Davidsdottir, Garnham, & Pauly, 1998; Khanna & Plassmann, 2004; Lantz & Feng, 2006; List & Gallet, 1999; Millimet, List, & Stengos, 2003; Panayotou, 1993; Selden & Song, 1994; Shafik, 1994; Song et al., 2013; Soytas, Sari, & Ewing, 2007; Taskin & Zaim, 2000), suspended particulate matter (SPM) (Binti Borhan & Musa Ahmed, 2010; Dinda, Coondoo, & Pal, 2000; Matsuoka, Matsumoto, & Kochi, 1998; Miah, Masum, Koike, Akther, & Muhammed, 2011; Selden & Song, 1994; Torras & Boyce, 1998; Vincent, 1997; Wheeler, 2001), respiratory suspended particulate matter (RSPM) (McConnell, 1997; Wheeler, 2001; Dasgupta, Laplante, Wang, & Wheeler, 2002; Rupasingha, Goetz, Debertin, & Pagoulatos, 2004; Alpay, 2005; Kumar & Foster, 2009; Orubu & Omotor, 2011; Feizpour & Shahmohammadi Mehrjardi, 2014), and several other pollutants as suitable proxies for environmental degradation. In view of this, it might be questioned the superiority of CO_2 as the predominant proxy for environmental degradation. Besides, we should not forget that CO_2 is considered as a global pollutant, whereas the aforementioned pollutants (e.g., NO_X, SO_2, SPM, RSPM) are considered as local pollutants. Therefore, while carrying out the EKC analysis for any singular geographic context, the local pollutants should be given priority, as the combustion of fossil fuel generates both the local and global pollutants. Rising level of CO_2 emissions is a global phenomenon, and therefore, the level of CO_2 emissions in a nation might not be the result of the industrial setup of that particular nation. Rather, the level of local pollutants can be directly attributed to the industrial structure of that nation.

Taking a cue from this discussion, we can suggest three procedures as possible solutions to this problem: First, while carrying out the EKC analysis for cross-country context, CO_2 emissions should be preferred over other pollutants. Second, while carrying out the EKC analysis for a single country, preference should be given to the local pollutants relevant for that country. Third, in continuation to the previous point, it is advisable to prepare an environmental degradation index, consisting of CO_2 and other local pollutants, and this can be carried out by the principal component analysis. These three procedures can solve the issue of choosing the proper proxy for environmental degradation in any geographic context. This will not only allow researchers to put forth the EKC analysis in a more comprehensive manner but also will provide more flexibility to the policy-makers in designing the suitable policies to reduce the local and global emission levels.

CONCLUSION

By far, we have discussed the problems faced by researchers in the course of an EKC analysis, and we have put forth some basic solutions to handle those problems. The problems were categorized under five broad categories, namely (a) model selection, (b) model validation, (c) data standardization, (d) variable selection and creation, and (e) proxy for environmental degradation. After a brief discussion of the problems, we have tried to provide the possible solutions to the problems using real data. The problems discussed in this chapter are mostly application oriented, and thus the solutions provided in this chapter are indicative, as with the course of time and with the more econometric development, researchers will be able to come up with better solutions to these problems.

Furthermore, to bring forth more sophistication and exclusivity in the EKC analysis, researchers might look beyond the variables already available in the public databases. Thereby, researchers might channelize their research agendas in the direction of sustainable development goals (SDGs), so that the multidimensional aspects can be brought under the purview of the EKC analysis. To achieve this, the specific aspects of the SDGs should be included as model variables in the EKC analysis. For example, EKC literature lacks in considering inequality variables, which are the pillars of SDG 10 (UNDP, 2017). If these variables are considered in the EKC models, then three more SDGs can be addressed in parallel, that is, (a) SDG 7—affordable and clean energy, (b) SDG 8—decent work and economic growth, and (c) SDG 13—climate action (UNDP, 2017). As the implementation of SDGs by 2030 is one of the major developmental agendas of the nations across the globe, following this framework in the EKC modeling can prove to be helpful for both the researchers and the policy-makers.

Keeping the solutions and the emerging research contexts in mind, the literature of EKC analysis is expected to be enhanced. While talking about the ecologic concerns, addressing these issues might result in larger socioeconomic implications in a global scale, thereby the results obtained from the EKC analysis might be used in a multidimensional way. The researchers should consider these aspects during the EKC analysis in any context, and then the relevance of EKC analysis can be established at a larger policy level.

REFERENCES

Abid, M. (2016). Impact of economic, financial, and institutional factors on CO_2 emissions: Evidence from Sub-Saharan Africa economies. *Utilities Policy, 41*, 85–94.

Abid, M. (2017). Does economic, financial and institutional developments matter for environmental quality? A comparative analysis of EU and MEA countries. *Journal of Environmental Management, 188*, 183–194.

Akbostancı, E., Türüt-Aşık, S., & Tunç, G.İ. (2009). The relationship between income and environment in Turkey: Is there an environmental Kuznets curve? *Energy Policy, 37*(3), 861–867.

Al Sayed, A. R., & Sek, S. K. (2013). Environmental Kuznets curve: Evidences from developed and developing economies. *Applied Mathematical Sciences, 7*(22), 1081–1092.

Alam, M. M., Murad, M. W., Noman, A. H. M., & Ozturk, I. (2016). Relationships among carbon emissions, economic growth, energy consumption and population growth: Testing Environmental Kuznets curve hypothesis for Brazil, China, India and Indonesia. *Ecological Indicators, 70*, 466–479.

Allard, A., Takman, J., Uddin, G. S., & Ahmed, A. (2018). The N-shaped environmental Kuznets curve: An empirical evaluation using a panel quantile regression approach. *Environmental Science and Pollution Research, 25*(6), 5848–5861.

Alpay, S. (2005). Interactions among economic development, openness to trade and environmental sustainability with a case study on South Korea. *The Journal of the Korean Economy, 6*(1), 67–89.

Álvarez-Herranz, A., Balsalobre-Lorente, D., Cantos, J. M., & Shahbaz, M. (2017a). Energy innovations-GHG emissions nexus: Fresh empirical evidence from OECD countries. *Energy Policy, 101*, 90–100.

Álvarez-Herranz, A., Balsalobre-Lorente, D., Shahbaz, M., & Cantos, J. M. (2017b). Energy innovation and renewable energy consumption in the correction of air pollution levels. *Energy Policy, 105*, 386–397.

Andreoni, J., & Levinson, A. (2001). The simple analytics of the environmental Kuznets curve. *Journal of Public Economics, 80*(2), 269–286.

Apergis, N., & Payne, J. E. (2009). Energy consumption and economic growth in Central America: Evidence from a panel cointegration and error correction model. *Energy Economics, 31*(2), 211–216.

Archibald, S. O., Banu, L. E., & Bochniarz, Z. (2004). Market liberalisation and sustainability in transition: Turning points and trends in central and Eastern Europe. *Environmental Politics, 13*(1), 266–289.

Bagliani, M., Bravo, G., & Dalmazzone, S. (2008). A consumption-based approach to environmental Kuznets curves using the ecological footprint indicator. *Ecological Economics, 65*(3), 650–661.

Balibey, M. (2015). Relationships among CO_2 emissions, economic growth and foreign direct investment and the EKC hypothesis in Turkey. *International Journal of Energy Economics and Policy, 5*(4), 1042–1049.

Balsalobre, D., & Alvarez, A. (2016). An approach to the effect of energy innovation on environmental Kuznets curve: An introduction to inflection point. *Bulletin of Energy Economics, 4*(3), 224−233.

Baltagi, B. H. (2005). *Econometric analysis of panel data.* Chichester, England: John Wiley & Sons.

van den Bergh, J. C., & Nijkamp, P. (1994). Dynamic macro modelling and materials balance: Economic-environmental integration for sustainable development. *Economic Modelling, 11*(3), 283−307.

Bimonte, S. (2002). Information access, income distribution, and the environmental Kuznets curve. *Ecological Economics, 41*(1), 145−156.

Bind Boilian, H., & Musa Ahmed, E. (2010). Pollution as one of the determinants of income in Malaysia: Comparison between single and simultaneous equation estimators of an EKC. *World Journal of Science, Technology and Sustainable Development, 7*(3), 291−308.

Biørn, E. (2004). Regression systems for unbalanced panel data: A stepwise maximum likelihood procedure. *Journal of Econometrics, 122*(2), 281−291.

Boamah, K. B., Du, J., Bediako, I. A., Boamah, A. J., Abdul-Rasheed, A. A., & Owusu, S. M. (2017). Carbon dioxide emission and economic growth of China—the role of international trade. *Environmental Science and Pollution Research, 24*(14), 13049−13067.

Bovenberg, A. L., & Smulders, S. (1995). Environmental quality and pollution-augmenting technological change in a two-sector endogenous growth model. *Journal of Public Economics, 57*(3), 369−391.

Brajer, V., Mead, R. W., & Xiao, F. (2008). Health benefits of tunneling through the Chinese environmental Kuznets curve (EKC). *Ecological Economics, 66*(4), 674−686.

Brajer, V., Mead, R. W., & Xiao, F. (2011). Searching for an environmental Kuznets curve in China's air pollution. *China Economic Review, 22*(3), 383−397.

Brown, S. P., & McDonough, I. K. (2016). Using the Environmental Kuznets curve to evaluate energy policy: Some practical considerations. *Energy Policy, 98*, 453−458.

Carson, R. T., Jeon, Y., & McCubbin, D. R. (1997). The relationship between air pollution emissions and income: US data. *Environment and Development Economics, 2*(4), 433−450.

Caviglia-Harris, J. L., Chambers, D., & Kahn, J. R. (2009). Taking the "U" out of Kuznets: A comprehensive analysis of the EKC and environmental degradation. *Ecological Economics, 68*(4), 1149−1159.

Cole, M. A., Rayner, A. J., & Bates, J. M. (1997). The environmental Kuznets curve: An empirical analysis. *Environment and Development Economics, 2*(4), 401−416.

Dasgupta, S., Laplante, B., & Mamingi, N. (2001). Pollution and capital markets in developing countries. *Journal of Environmental Economics and Management, 42*(3), 310−335.

Dasgupta, S., Laplante, B., Wang, H., & Wheeler, D. (2002). Confronting the environmental Kuznets curve. *The Journal of Economic Perspectives, 16*(1), 147−168.

De Bruyn, S. M., van den Bergh, J. C., & Opschoor, J. B. (1998). Economic growth and emissions: Reconsidering the empirical basis of environmental Kuznets curves. *Ecological Economics, 25*(2), 161−175.

Dijkgraaf, E., & Vollebergh, H. R. (2005). A test for parameter homogeneity in CO_2 panel EKC estimations. *Environmental and Resource Economics, 32*(2), 229−239.

Dinda, S., Coondoo, D., & Pal, M. (2000). Air quality and economic growth: An empirical study. *Ecological Economics, 34*(3), 409−423.

Dong, B., Wang, F., & Guo, Y. (2016). The global EKCs. *International Review of Economics & Finance, 43*, 210−221.

Efron, B., Hastie, T., Johnstone, I., & Tibshirani, R. (2004). Least angle regression. *The Annals of Statistics, 32*(2), 407−499.

Egli, H. (2001). *Are cross-country studies of the environmental Kuznets curve misleading? New evidence from time series data for Germany.* Ernst Moritz Arndt University of Greifswald, Faculty of Law and Economics. Working paper no. 10/2001.

Ercolano, S., Gaeta, G. L. L., Ghinoi, S., & Silvestri, F. (2018). Kuznets curve in municipal solid waste production: An empirical analysis based on municipal-level panel data from the Lombardy region (Italy). *Ecological Indicators, 93*, 397−403.

Feizpour, M. A., & Shahmohammadi Mehrjardi, A. (2014). Comparative advantages and sustainable development in Iranian manufacturing industries. *Iranian Economic Review, 18*(1), 33−52.

Fonkych, K., & Lempert, R. (2005). Assessment of environmental Kuznets curves and socioeconomic drivers in IPCC's SRES scenarios. *The Journal of Environment & Development, 14*(1), 27−47.

Galeotti, M., & Lanza, A. (2005). Desperately seeking environmental Kuznets. *Environmental Modelling & Software, 20*(11), 1379−1388.

Galeotti, M., Lanza, A., & Pauli, F. (2006). Reassessing the environmental Kuznets curve for CO_2 emissions: A robustness exercise. *Ecological Economics, 57*(1), 152−163.

Gradus, R., & Smulders, S. (1993). The trade-off between environmental care and long-term growth—pollution in three prototype growth models. *Journal of Economics, 58*(1), 25−51.

Grossman, G. M., & Krueger, A. B. (1991). *Environmental impacts of a North American free trade agreement.* National Bureau of Economic Research. Working paper no. w3914.

Grossman, G. M., & Krueger, A. B. (1995). Economic growth and the environment. *The Quarterly Journal of Economics, 110*(2), 353−377.

Halkos, G. E. (2003). Environmental Kuznets curve for sulfur: Evidence using GMM estimation and random coefficient panel data models. *Environment and Development Economics, 8*(4), 581−601.

Harbaugh, W. T., Levinson, A., & Wilson, D. M. (2002). Reexamining the empirical evidence for an environmental Kuznets curve. *The Review of Economics and Statistics, 84*(3), 541−551.

He, J., & Richard, P. (2010). Environmental Kuznets curve for CO_2 in Canada. *Ecological Economics, 69*(5), 1083–1093.

Hettige, H., Mani, M., & Wheeler, D. (2000). Industrial pollution in economic development: The environmental Kuznets curve revisited. *Journal of Development Economics, 62*(2), 445–476.

Hill, R. J., & Magnani, E. (2002). An exploration of the conceptual and empirical basis of the environmental Kuznets curve. *Australian Economic Papers, 41*(2), 239–254.

Jones, L., & Manuelli, R. (1995). *A positive model of growth and pollution controls (No. w5205).* National Bureau of Economic Research.

Kaufmann, R. K., Davidsdottir, B., Garnham, S., & Pauly, P. (1998). The determinants of atmospheric SO_2 concentrations: Reconsidering the environmental Kuznets curve. *Ecological Economics, 25*(2), 209–220.

Khanna, N., & Plassmann, F. (2004). The demand for environmental quality and the environmental Kuznets curve hypothesis. *Ecological Economics, 51*(3), 225–236.

Kumar, N., & Foster, A. D. (2009). Air quality interventions and spatial dynamics of air pollution in Delhi and its surroundings. *International Journal of Environment and Waste Management, 4*(1–2), 85–111.

Kuznets, S. (1955). Economic growth and income inequality. *The American Economic Review, 45*(1), 1–28.

Lantz, V., & Feng, Q. (2006). Assessing income, population, and technology impacts on CO_2 emissions in Canada: Where's the EKC? *Ecological Economics, 57*(2), 229–238.

Lee, C. C., Chiu, Y. B., & Sun, C. H. (2009). Does one size fit all? A reexamination of the environmental Kuznets curve using the dynamic panel data approach. *Review of Agricultural Economics, 31*(4), 751–778.

Lipford, J. W., & Yandle, B. (2010). Environmental Kuznets curves, carbon emissions, and public choice. *Environment and Development Economics, 15*(04), 417–438.

List, J. A., & Gallet, C. A. (1999). The environmental Kuznets curve: Does one size fit all? *Ecological Economics, 31*(3), 409–423.

Martínez-Zarzoso, I., & Bengochea-Morancho, A. (2004). Pooled mean group estimation of an environmental Kuznets curve for CO_2. *Economics Letters, 82*(1), 121–126.

Matsuoka, S., Matsumoto, R., & Kochi, I. (1998). Economic growth and environmental quality in developing countries: A verification of environmental Kuznets curve. *Environmental Sciences, 11*(4), 349–362.

Mazzanti, M., Montini, A., & Zoboli, R. (2008). Environmental Kuznets curves for air pollutant emissions in Italy: Evidence from environmental accounts (NAMEA) panel data. *Economic Systems Research, 20*(3), 277–301.

McConnell, K. E. (1997). Income and the demand for environmental quality. *Environment and Development Economics, 2*(4), 383–399.

Miah, M. D., Masum, M. F. H., & Koike, M. (2010). Global observation of EKC hypothesis for CO_2, SO_x and NO_x emission: A policy understanding for climate change mitigation in Bangladesh. *Energy Policy, 38*(8), 4643–4651.

Miah, M. D., Masum, M. F. H., Koike, M., Akther, S., & Muhammed, N. (2011). Environmental Kuznets curve: The case of Bangladesh for waste emission and suspended particulate matter. *The Environmentalist, 31*(1), 59–66.

Millimet, D. L., List, J. A., & Stengos, T. (2003). The environmental Kuznets curve: Real progress or misspecified models? *The Review of Economics and Statistics, 85*(4), 1038–1047.

Mohapatra, G., & Giri, A. K. (2009). Economic development and environmental quality: An econometric study in India. *Management of Environmental Quality: An International Journal, 20*(2), 175–191.

Moomaw, W. R., & Unruh, G. C. (1997). Are environmental Kuznets curves misleading us? The case of CO_2 emissions. *Environment and Development Economics, 2*(04), 451–463.

Narayan, P. K., & Narayan, S. (2010). Carbon dioxide emissions and economic growth: Panel data evidence from developing countries. *Energy Policy, 38*(1), 661–666.

das Neves Almeida, T. A., Cruz, L., Barata, E., & García-Sánchez, I. M. (2017). Economic growth and environmental impacts: An analysis based on a composite index of environmental damage. *Ecological Indicators, 76,* 119–130.

Onafowora, O. A., & Owoye, O. (2014). Bounds testing approach to analysis of the environment Kuznets curve hypothesis. *Energy Economics, 44,* 47–62.

Orubu, C. O., & Omotor, D. G. (2011). Environmental quality and economic growth: Searching for environmental Kuznets curves for air and water pollutants in Africa. *Energy Policy, 39*(7), 4178–4188.

Pal, D., & Mitra, S. K. (2017). The environmental Kuznets curve for carbon dioxide in India and China: Growth and pollution at crossroad. *Journal of Policy Modeling, 39*(2), 371–385.

Panayotou, T. (1993). *Empirical tests and policy analysis of environmental degradation at different stages of economic development.* International Labour Organization. Working paper no. 292778.

Pao, H. T., & Tsai, C. M. (2010). CO_2 emissions, energy consumption and economic growth in BRIC countries. *Energy Policy, 38*(12), 7850–7860.

Pao, H. T., & Tsai, C. M. (2011). Multivariate granger causality between CO_2 emissions, energy consumption, FDI (foreign direct investment) and GDP (gross domestic product): Evidence from a panel of BRIC (Brazil, Russian federation, India, and China) countries. *Energy, 36*(1), 685–693.

Papież, M., Śmiech, S., & Frodyma, K. (2018). Determinants of renewable energy development in the EU countries. A 20-year perspective. *Renewable and Sustainable Energy Reviews, 91,* 918–934.

Park, S., & Lee, Y. (2011). Regional model of EKC for air pollution: Evidence from the Republic of Korea. *Energy Policy, 39*(10), 5840–5849.

Richmond, A. K., & Kaufmann, R. K. (2006). Is there a turning point in the relationship between income and energy use and/or carbon emissions? *Ecological Economics, 56*(2), 176–189.

Roberts, J. T., & Grimes, P. E. (1997). Carbon intensity and economic development 1962–1991: A brief exploration of the environmental Kuznets curve. *World Development, 25*(2), 191–198.

Roca, J., & Alcántara, V. (2001). Energy intensity, CO_2 emissions and the environmental Kuznets curve. The Spanish case. *Energy Policy, 29*(7), 553–556.

Roumasset, J., Burnett, K., & Wang, H. (2006). *Environmental resources and economic growth*. Washington, DC: World Bank.

Rupasingha, A., Goetz, S. J., Debertin, D. L., & Pagoulatos, A. (2004). The environmental Kuznets curve for US counties: A spatial econometric analysis with extensions. *Papers in Regional Science, 83*(2), 407–424.

Sebri, M. (2016). Testing the environmental Kuznets curve hypothesis for water footprint indicator: A cross-sectional study. *Journal of Environmental Planning and Management, 59*(11), 1933–1956.

Selden, T. M., & Song, D. (1994). Environmental quality and development: Is there a Kuznets curve for air pollution emissions? *Journal of Environmental Economics and Management, 27*(2), 147–162.

Shafik, N. (1994). Economic development and environmental quality: An econometric analysis. *Oxford Economic Papers, 46*(4), 757–773.

Shahbaz, M., Balsalobre, D., & Shahzad, S. J. H. (2018a). The influencing factors of CO_2 emissions and the role of biomass energy consumption: Statistical experience from G-7 countries. *Environmental Modeling & Assessment*, 1–19.

Shahbaz, M., Hye, Q. M. A., Tiwari, A. K., & Leitão, N. C. (2013a). Economic growth, energy consumption, financial development, international trade and CO_2 emissions in Indonesia. *Renewable and Sustainable Energy Reviews, 25*, 109–121.

Shahbaz, M., Lean, H. H., & Shabbir, M. S. (2012). Environmental Kuznets curve hypothesis in Pakistan: Cointegration and granger causality. *Renewable and Sustainable Energy Reviews, 16*(5), 2947–2953.

Shahbaz, M., Loganathan, N., Muzaffar, A. T., Ahmed, K., & Jabran, M. A. (2016a). How urbanization affects CO_2 emissions in Malaysia? The application of STIRPAT model. *Renewable and Sustainable Energy Reviews, 57*, 83–93.

Shahbaz, M., Mahalik, M. K., Shah, S. H., & Sato, J. R. (2016b). Time-varying analysis of CO_2 emissions, energy consumption, and economic growth nexus: Statistical experience in next 11 countries. *Energy Policy, 98*, 33–48.

Shahbaz, M., Nasir, M. A., & Roubaud, D. (2018b). Environmental degradation in France: The effects of FDI, financial development, and energy innovations. *Energy Economics, 74*, 843–857.

Shahbaz, M., Ozturk, I., Afza, T., & Ali, A. (2013b). Revisiting the environmental Kuznets curve in a global economy. *Renewable and Sustainable Energy Reviews, 25*, 494–502.

Shahbaz, M., & Sinha, A. (2019). Environmental Kuznets curve for CO2 emissions: a literature survey. *Journal of Economic Studies, 46*(1), 106–168.

Shahbaz, M., Solarin, S. A., & Ozturk, I. (2016c). Environmental Kuznets curve hypothesis and the role of globalization in selected African countries. *Ecological Indicators, 67*, 623–636.

Sinha, A. (2015). Modeling energy efficiency and economic growth: Evidences from India. *International Journal of Energy Economics and Policy, 5*(1), 96–104.

Sinha, A. (2016). Trilateral association between SO_2/NO_2 emission, inequality in energy intensity, and economic growth: A case of Indian cities. *Atmospheric Pollution Research, 7*(4), 647–658.

Sinha Babu, S., & Datta, S. K. (2013). The relevance of environmental Kuznets curve (EKC) in a framework of broad-based environmental degradation and modified measure of growth–a pooled data analysis. *The International Journal of Sustainable Development and World Ecology, 20*(4), 309–316.

Sinha, A., & Bhattacharya, J. (2016). Environmental Kuznets curve estimation for NO_2 emission: A case of Indian cities. *Ecological Indicators, 67*, 1–11.

Sinha, A., & Bhattacharya, J. (2017). Estimation of environmental Kuznets curve for SO_2 emission: A case of Indian cities. *Ecological Indicators, 72*, 881–894.

Sinha, A., & Rastogi, S. K. (2017). Collaboration between central and state government and environmental quality: Evidences from Indian cities. *Atmospheric Pollution Research, 8*(2), 285–296.

Sinha, A., & Shahbaz, M. (2018). Estimation of environmental Kuznets curve for CO_2 emission: Role of renewable energy generation in India. *Renewable Energy, 119*, 703–711.

Sinha, A., Shahbaz, M., & Balsalobre, D. (2017). Exploring the relationship between energy usage segregation and environmental degradation in N-11 countries. *Journal of Cleaner Production, 168*, 1217–1229.

Song, M. L., Zhang, W., & Wang, S. H. (2013). Inflection point of environmental Kuznets curve in Mainland China. *Energy Policy, 57*, 14–20.

Soytas, U., Sari, R., & Ewing, B. T. (2007). Energy consumption, income, and carbon emissions in the United States. *Ecological Economics, 62*(3), 482–489.

Stern, D. I. (2004). The rise and fall of the environmental Kuznets curve. *World Development, 32*(8), 1419–1439.

Tamazian, A., Chousa, J. P., & Vadlamannati, K. C. (2009). Does higher economic and financial development lead to environmental degradation: Evidence from BRIC countries. *Energy Policy, 37*(1), 246–253.

Tan, J., Liu, H., Li, M., & Wang, J. (2018). A prediction scheme of tropical cyclone frequency based on lasso and random forest. *Theoretical and Applied Climatology, 133*(3–4), 973–983.

Taskin, F., & Zaim, O. (2000). Searching for a Kuznets curve in environmental efficiency using kernel estimation. *Economics Letters, 68*(2), 217–223.

Torras, M., & Boyce, J. K. (1998). Income, inequality, and pollution: A reassessment of the environmental Kuznets curve. *Ecological Economics, 25*(2), 147–160.

Uddin, G. A., Alam, K., & Gow, J. (2016). Does ecological footprint impede economic growth? An empirical analysis based on the environmental Kuznets curve hypothesis. *Australian Economic Papers, 55*(3), 301–316.

United Nations Development Programme (UNDP). (2017). *Sustainable development Goals*. Available at: http://www.undp.org/content/undp/en/home/sustainable-development-goals.html.

Ver Hoef, J. M., Peterson, E. E., Hooten, M. B., Hanks, E. M., & Fortin, M. J. (2018). Spatial autoregressive models for statistical inference from ecological data. *Ecological Monographs, 88*(1), 36–59.

Verdier, T. (1993). *Environmental pollution and endogenous growth A comparison between emission taxes and technological standards.* Fondazione Eni Enrico Mattei. Discussion Paper 57.

Vincent, J. R. (1997). Testing for environmental Kuznets curves within a developing country. *Environment and Development Economics, 2*(04), 417–431.

Wang, Y., Zhang, X., Kubota, J., Zhu, X., & Lu, G. (2015). A semi-parametric panel data analysis on the urbanization-carbon emissions nexus for OECD countries. *Renewable and Sustainable Energy Reviews, 48,* 704–709.

Welsch, H. (2004). Corruption, growth, and the environment: A cross-country analysis. *Environment and Development Economics, 9*(5), 663–693.

Wheeler, D. (2001). Racing to the bottom? Foreign investment and air pollution in developing countries. *The Journal of Environment & Development, 10*(3), 225–245.

CHAPTER 8

Time Series Analysis (Stationarity, Cointegration, and Causality)

ABDUL JALIL, PHD • NASIR HAMID RAO, MPHIL SCHOLAR

The environmental Kuznets curve (EKC) hypothesis is widely tested by applied econometricians over the last 25 years. The literature presents plenty of studies, which are based on cross-sectional, pooled, and panel data. Besides this, the testing of EKC hypothesis is also popularized on the country cases over the last 10 years. Jalil and Mahmud (2009), one of the earliest and most cited studies in the time-series literature on EKC, test the case of China in the framework of cointegration. Then there is a plethora of research that deals with individual country cases in the context of EKC hypothesis.[1] However, most of the time if not all, the mushrooming number of articles and studies based on the time series analysis are inappropriate in the context of EKC. Consequently, the policy recommendations, by inappropriate analyses, may misguide the stakeholder. Therefore, it is important to understand the time series literature in an appropriate and applied sense. This chapter deals with the same objective.

Time series econometrics mainly deals with careful collection and rigorous study of the past observations of variables with an aim to develop appropriate models, which can potentially describe the integral structure of the data series. This branch of econometrics may be divided into two main categories, that is, forecasting and dynamic modeling. Generally, researchers do not test any economic theory or hypothesis in the case of forecasting but try to understand the structure of time series data on the basis of different characteristics of the series and then try to forecast. Such models are then used to make forecasts, which are essential for forward-looking policy-making. This may be univariate analysis or multivariate analysis. On the other hand, the *causation* and the relationships, both in the short run and the long run, are investigated in dynamic modeling.

In this backdrop, we may think that there are a number of concepts in time series analysis, which require the comprehensive understanding, for example, stationarity, nonstationarity, causation, cointegration, and error correction models. Therefore, this chapter will explain the concepts of stationarity, cointegration, and causality in the framework of time series analysis along with some examples. Thereby, researchers will have knowledge about the econometric tools and methods used in the single-country studies in the EKC literature. The practical importance of these methods will be put across in simple words. Finally, the chapter will discuss the major pros and cons of various methods. For instance, how does the researcher decide between a forecasting model and a dynamic modeling approach? However, before moving on, it is important to grasp on the components of time series.

Rest of the chapter is organized as follows. Section Components of a Time Series explains the basic components of time series. The time series analysis is described in Section Analysis of Time Series Data Series. The concept of stationarity and nonstationarity is discussed in detail in Section Stationary and Nonstationary Time Series. There are various quantitative tests for the testing of stationarity properties of the data series, which are described in Section Tests of Stationarity. The cointegration analysis is discussed in Sections Unit Root Tests with Structural Breaks and Cointegration and Error Correction Model Without Structural Breaks concludes the chapter.

---

[1] See for example Acaravci and Ozturk (2010), Lipford and Yandle (2010), Jalil and Feridun (2011), Pao, Yu, and Yang (2011), Esteve and Tamarit (2012a), Esteve and Tamarit (2012b), Abdullah et al. (2012), Ozcan (2013), Ozturk and Acaravci (2013), Cho et al. (2014), Lav et al. (2014), Onafowora and Owoyes (2014) Al Shehry (2015), Farhani and Ozturk (2015), Ozturk and Al-Mulali (2015), Ahmad et al. (2016), Dogan and Turkekul (2016), Ertugrul et al. (2016), Ozatac et al. (2017) and Yavuz (2014) among many others.

Environmental Kuznets Curve (EKC). https://doi.org/10.1016/B978-0-12-816797-7.00008-4

COMPONENTS OF A TIME SERIES

Generally, an observed time series has four main components. These components are trend, cyclic, seasonal, and irregular (TSCI). It is important to mention that each component recommends different breed of econometric models. Therefore, a detailed analysis of these components is utmost important. More specifically, every time series potentially has different strength of these factors; therefore, any wrong judgment about the disaggregated source could lead to a disaster of a forecast from a time series model. Hence, this is really important to understand the dimensions of TSCI components before building a time series model.

Trend: If time series data have a general trend, then it implies that the time series has a general *trend* of increasing, decreasing, or stagnation over a long period of time. The prime examples of trend are the increasing appetite of gross domestic product (GDP) of an economy, increasing trend in CO_2 emissions due to use of fossil fuels, decreasing trend of life-threatening diseases because of medical advancements and persistence of inflation in developing countries. Trend is really important while building models to make forecasts. There are several ways, ranging from visual analysis such as line chart to econometric analysis such as regression-based trend model, available to researcher to identify and determine the trend of a series.

Seasonal: The variation that occurs in a time series data due to *season-to-season* change is termed as *seasonal*. Seasonal pattern is the one, which is repeatedly happening in the time series, but at very specific part of the year. The important factors that contribute in seasonal variations could be the change of weather conditions, customs and traditions, etc. The different sales patterns of companies in different quarters could be the example of seasonality. There are several ways, ranging from plotting simple charts to theoretic econometrics models, to identify and gauging the strength of seasonal patterns in a time series.

Cyclic: This component of time series explains the medium-term variations in the series, caused by circumstances but repeated in cycles. Duration of cycle varies with the kind of time series; generally it extends over longer period of time, usually two or more years. Identification of cyclic patterns can be done either by using deterministic econometric models or by using prior information on cycles. The prime examples of cyclic variations are the inflationary trends in different parts of economic cycle and growth trajectories across cycles.

Irregular: These are the random variations in a data series, which are caused by unpredictable influences. These variations do not repeat in a particular pattern. This is the part that generally makes the time series a stochastic process.

ANALYSIS OF TIME SERIES DATA SERIES

Generally, the time series analysis may be divided into main branches, that is, forecasting and dynamic modeling. The time series forecasting means that predicting the future values based on the understanding of previous data distribution, that is, mean, variance, skewness, and kurtosis. It implies that we do not test any type of hypothesis or economic theory in this branch of time series analysis. We simply try to understand the patterns of the time series through different statistics and predict the future. More clearly, generally, we are not concerned with hypothesis testing, investigating the relationships or building models in this line of research. We just try to extract the information from the past to predict the future. On the other, in dynamic modeling, we build econometric models to investigate the relationship, test the hypothesis, and validate the theory.

The time series forecasting can further be divided into univariate analysis and multivariate analysis. The univariate analysis was popularized by Box and Jenkins (1976) by introducing the data-generating process (DGP) of autoregressive (AR), moving average (MA), and ARMA. Then Engel (1982) highlights the issue of conditional heteroscedasticity in autoregressive conditional heteroscedasticity (ARCH) model, which are extended in various forms, such as generalized ARCH (GARCH), exponential GARCH (EGARCH), and threshold ARCH (TARCH) by various econometricians. However, univariate forecasting is not useful in the context of EKC. Therefore, we shall not discuss it in detail. Then multivariate models are introduced and popularized by Sims (1980) by introducing vector autoregressive (VAR) models. Generally, Sims (1980) criticizes the demarcation between exogenous and endogenous variables and postulates that everything is causing everything. Because of this extreme assumption, the economists do not consider it useful for establishing the long-run relationships among the variables. However, the Granger causality test and impulse response functions are the best outcomes of VAR analysis in the context of EKC.

The other branch of time series analysis is dynamic modeling. This branch is more relevant in the context of EKC. Particularly, we test hypothesis, investigate economic relationships, and validate economic theories in this branch of time series. There are two main streams in

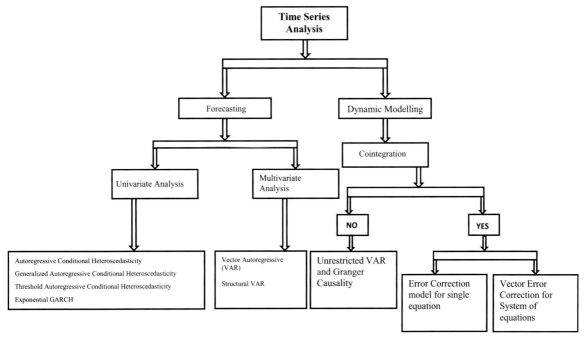

FIG. 8.1 The basic framework of time series analysis.

this branch, based on the decision whether there exist a long-run relationship or not. If there exist a long-run relationship among the underline variables, then we move to cointegration analysis and *error correction models*. If there is no long-run relationship among the variables, then rely on the Granger causality and impulse response function for discussing the economic relationships (Fig. 8.1).

STATIONARY AND NONSTATIONARY TIME SERIES

It is argued that the natural start of the time series analysis is testing the stationarity property of the time series data. In simple regression models, for example ordinary least square (OLS) models, there is a huge assumption that the data series is stationary. However, there is a great possibility, in most of the cases, of a nonstationary series in the case of economic variables. Indeed, it is well settled in the time series literature that there may be a spurious regression in the case of nonstationary data series. Therefore, testing the stationarity is the basic step in both forecasting and dynamic modeling analysis.

Concept of Stationarity

The field of time series is a vast one that encompasses many areas of study and concepts. The EKC is also

one of them. Most of the literature on EKC uses the data series of CO_2 emission and GDP to analyze the existence of EKC. As mentioned earlier, the natural start of time series analysis is to test stationarity properties. Additionally, most of the econometric models concentrate mainly on the stationary property of the data for a meaningful model. Therefore, the standard books on time series analysis start with the concept of stationarity. However, it will be quite easy to understand the concept of stationarity if one can remember the very basic definition of the parameters estimated through any econometric model. It is important to remember that parameters explain "on average" relations between dependent and independent variables. Hence, to have meaningful results from an econometric model, it is essential to determine whether the data are stationary or not.

By stationarity, we mean that data are mean reverting. There are a lot of econometric books that have defined stationary series as the one having statistical properties such a mean, variance, and covariance as constant. However, practically, there is substantial number of instances where mean of the variable is not varying, but overall data have the tendency to revert to their mean. In such scenarios, time series is stationary despite the fact that mean of the series is nonconstant. Hence, we can conclude that concept of stationarity is mainly connected with the reversion of series to its mean instead of putting this in the context of constant mean or variance.

The Importance of Stationarity

Should we worry about the stationarity property? Yes, there are several reasons. First, if the time series is not stationary, then the study of the series is meaningful only for the underconsideration time period. The behavior cannot be generalized for the other time series due to its non-mean-reversion property. Therefore, the nonstationary time series will not give meaningful forecast. Second, as mentioned earlier, the concept of stationarity is really important in building an effective and correct econometric model. Just imagine we run a regression for Y by taking a variable X as its determinant. If regression is estimated through OLS method, we will get one intercept (alpha) and one slope coefficient (beta). Beta will potentially be explaining "how much one unit on average change in X can bring a change on average value of Y." Now if X and Y are showing a non-mean-reverting behavior, then we are technically relying on a wrong model. As model is telling us about the relationship between average values of X and Y while both variables have proven tendency of nonreversion to mean, technically model is explaining nothing, and any conclusions drawn out of that model will be misleading. The model will be explaining on average relationship, whereas the actual series have tendency not to revert to their means. This is the very reason that stationarity is really important.

TESTS OF STATIONARITY

There are several tests to test the stationarity properties in the time series data. If we want to get a first guess about the time series, then we shall plot the data series. It will give an initial clue and intuitive feel about the time series data. It is worth mentioning that plotting the data is important as well. Chatfield (2004) notes that "Anyone who tries to analyze a time series without plotting it first is asking for trouble." The second, well discussed in the time series literature, is autocorrelation function (ACF) and correlogram. The ACF is generally used for the ARMA and ARCH/GARCH framework. Therefore, we shall not give the details here. We shall concentrate on the statistical test, which is more useful in the context of EKC.

Dickey-Fuller Test

The Dickey-Fuller (DF) test was developed and popularized by Dickey and Fuller (1979). The null hypothesis

of DF test is that there is a unit root in an AR model, which implies that the data series is not stationary. The alternative hypothesis is generally stationarity or trend stationarity but can be different depending on the version of the test is being used. Consider the following example to understand DF test.

$$co_{2t} = \rho co_{2t-1} + \mu_t \qquad (8.1)$$

The equation represents the first-order AR (AR[1]) process, implies that the co_2 (a measure of environmental degradation) depends on its past value only, that is, the variable is explained by its own previous lag. The ρ is the coefficient of interest for the determination of unit root or stationarity. If value of ρ is less than one, shocks in co_2 will be transitory and will die out with the passage of time. In this case, the series will be stationary in nature. Alternatively, if the value of ρ is equal to one, shocks in co_2 data series are permanent in nature and do not die out with the passage of time. In such scenario, the series is characterized as unit root. If it is greater than one, then series will explode. Let us see the figure for all three cases.

The problem with this equation is that the upfront assumption is the presence of stationarity, which is the very reason that equation has been estimated at levels. However, theory suggests that the starting point should be the doubts, which we want to clear. Here the doubts are that the series is unit root; hence, we should incorporate these doubts in the testing model. Therefore, we can rewrite the model by the first difference, and the model will be as follows:

$$\Delta co_{2t} = (\rho - 1)co_{2t-1} + u_t = \delta co_{t-1} + u_t \qquad (8.2)$$

Here the upfront assumption is that the series is nonstationary and that is why the model is estimated in different form. Here δ is the coefficient of interest for determination of unit root or stationarity. If value of δ is less than zero, shocks in co_2 will be transitory and will die out with the passage of time. In this case, the series will be stationary in nature. Alternatively, if the value of δ is equal to zero, shocks in co_2 are permanent in nature and do not die out with the passage of time. In such scenario, the series is characterized as unit root. This is called unit root hypothesis. One can argue that this model has ignored the trend and intercept and hence the coefficient estimated could not be true. This is a very logical criticism and needs a careful review. We can add intercept and trend, and the models could be

$$\Delta co_{2t} = \alpha_0 + \delta co_{2t-1} + u_t \quad \text{(with drift) or} \qquad (8.3)$$

$$\Delta co_{2t} = \alpha_0 + \alpha_1 t + \delta co_{2t-1} + u_t \quad \text{(with drift and trend)} \qquad (8.4)$$

However, this is not a straightforward decision to choose from these three models. Importantly, the decision is important because the inclusion or exclusion of intercept or trend term may lead to biased results. Therefore, better approach is to decide on whether to include or exclude intercept and trend term could be based on the deep analysis of the data generation process of the series. Alternatively, OLS regression can also be used to make a decision on the inclusion or exclusion.

Furthermore, DF test is treated as one of the weakest tests for the presence of unit root, as it only incorporates the AR (1) DGP for the series. This is a really tough and stringent assumption. As most of the time, series do carry a long-term association with its historic events. Furthermore, the above can carry the higher-order autocorrelation as well. Therefore, there may be two alternative approaches to handle the autocorrelation. These are parametric approach and nonparametric approach. The alternative and better parametric test is augmented Dickey-Fuller (ADF) test, and the nonparametric test is Philip and Perron (1988) test.

Augmented Dickey-Fuller Test

As mentioned earlier, the null hypothesis of DF test is that a unit root is present in a first-order AR model. The first-order AR process captures the first-order autocorrelation as well. If we expect a higher-order autocorrelation in the model, then DF is failed, and white noise assumption of error term will be violated. Therefore, Dicky and Fuller augment the equation with higher-order lags to capture the higher-order autocorrelation, which in known as Augmented Dickey Fuller (ADF) test in the literature of time series. More clearly, ADF test is used to parametrically correct the higher-order correlation by taking the higher number of lags on the right-hand side of the equation.

$$\Delta co_t + \alpha_0 + \alpha_1 t + \delta co_{t-1} + \beta_1 \Delta co_{t-1} + \beta_2 \Delta co_{t-2}$$
$$+ \ldots\ldots\ldots \beta_p \Delta co_{t-p} + u_t \qquad (8.5)$$

Inclusion of number of lags on the independent side is a practical approach in the above equation. Generally, the applied econometricians advice that one should include sufficient number of lags to remove autocorrelation in the residuals. However, sometime it is not particularly useful due to the loss of degree of freedom. The other major limitation of ADF is that it does not incorporate the structural changes and shifts in the data under study.

Phillips-Perron Test

The nonparametric methodology to handle the issue autocorrelation, when testing for a unit root, is

presented by Phillips and Perron (1988). Therefore, it is known Phillips-Perron test. We estimate nonaugmented DF equation to test the unit root and then modify the inferences of the coefficients to correct the impacts of autocorrelations. Phillips-Perron test makes a nonparametric correction to the t-test statistic and provides more robust results with respect to unspecified autocorrelation and heteroscedasticity in the disturbance process of the test equation. Most of the times, this test is used when there is a series loss of degree of freedom due to small number of observation with higher-order autocorrelations.

UNIT ROOT TESTS WITH STRUCTURAL BREAKS

Indeed, DF, ADF, and Phillip-Perron tests are well known and standard tests to detect the unit roots in the time series data. However, the recent literature of time series analysis argues that these tests have the low power to test the unit root hypothesis in the presence of structural brakes, structural changes, or sudden changes in underlying time series data. More clearly, if we fail to account the structural breaks and structural changes, then the conventional tests may produce the misleading results about the inferences. Specifically, Perron (1989) notes that the conventional unit root tests, such as ADF and Phillips-Perron tests, may produce incorrect inference in the presence of structural breaks.

Therefore, we need to incorporate the structural breaks during testing of the unit root test. There are several possibilities, which are discussed by the researchers while testing the unit root hypothesis. For example, Zivot and Andrews (1992) single break in the data series, Lumsdaine and Papell (1997) discuss more than one break. Clemente, Montanes, and Reyes (1998) incorporate two types of breaks, that is, innovative outliers (IOs) and additive outliers (AOs). This specific section deals with unit root testing with structural breaks (Box 8.1).

Zivot and Andrews (1992) Unit Root Test

Perron (1989) suggests that ADF test can be modified by including a known structural break or exogenous structural break. Then Zivot and Andrews (1992) test determines the break point endogenously from the time-time series data. Specially, there are three different models, proposed by Zivot and Andrews (1992), to test unit root hypothesis in the presence of one structural break.

$$\Delta co_t = \mu + \alpha co_{t-1} + \beta t + \theta_1 du_t + \sum_{j=1}^{k} dj \Delta co_{t-j} + \varepsilon_t \qquad (8.6)$$

BOX 8.1
Structural Breaks

The structural break, structural change or sudden change in the DGP, is a very well-known term in the literature of time series. Generally, the concept of structural break implies that unexpected shift in a time series may lead to disaster errors in the forecasting. The forecasting models become unreliable in the presence of structural breaks. Consequently, it is argued that an applied econometrician must routinely test the stability or structural breaks in the models. There are several tests that are proposed to test the structural breaks. Chow test, proposed and popularized by Chow (xxx), is very famous in the time series literature to test the structural breaks. However, it is designed for the case of single break in the mean with a known time period. Afterward, it is realized that there are many other cases such as more than one break with known time period, single break with unknown time period, and more than one break with unknown time period may arise in a DGP. Therefore, Chow test becomes an appropriate choice in these cases. Resultantly, a number of tests for detecting the number of breaks are introduced by the econometricians.

It is also important to mention here that the conventional unit root tests, such as DF, ADF, Phillips-Perron test, are not applicable in the presence of structural breaks. Therefore, we have to use the other important streams of tests, which incorporate various types of structural breaks. For example, Zivot and Andrew test for one endogenous break in mean and single break in trend or Lumsdaine-Papell test for two breaks in mean and two breaks in trend. Clemente et al. (1998) proposed another unit root test for two structural breaks. Importantly, Clemente-Montanes-Reyes unit root test produces the results of two different types of structural breaks. First, AO implies a sudden change, and IO implies the gradual change.

Similarly, the cointegration methods such as Engel-Granger cointegration test, Johansen-Juselius cointegration test, and autoregressive distributed lag cointegration are not applicable in the presence of structural breaks. Therefore, we use Gregory and Hansen (1996) cointegration tests for the presence of structural break at an unknown time period. We estimate three different models in this regard, that is, shift in mean, shift in trend, and shift in both mean and trend. The detail is presented in the main text.

$$\Delta co_t = \mu + \alpha co_{t-1} + \beta t + \gamma_1 dt_t + \sum_{j=1}^{k} dj\Delta co_{t-j} + \varepsilon_t \quad (8.7)$$

$$\Delta co_t = \mu + \alpha co_{t-1} + \beta t + \theta_1 du_t + \gamma_1 dt_t + \sum_{j=1}^{k} dj\Delta co_{t-j} + \varepsilon_t \quad (8.8)$$

Here, k represents the number of lags or augmentation order, du_t is the dummy variable for a mean shift occurring at time of break (TB), and dt_t is the corresponding trend shift variable.

It is evident from Eq. (8.6) that structural break is in the intercept term, that is, du_t. The structural break in trend is shown in Eq. (8.7), and structural breaks in both, that is intercept and trend, are shown in Eq. (8.8). We need a grid search procedure to implement Zivot and Andrews unit root test for testing the unit root hypothesis. This is a procedure that starts from estimating the basic equation through OLS and then test the unit root with null hypothesis $\alpha = 0$ for each potential structural break.

Lumsdaine and Papell (1997) Unit Root Test

The major drawback of Zivot and Andrews (1992) is that it is designed only for one break. Therefore, Lumsdaine and Papell (1997) extend the procedure of

Zivot and Andrews (1992) for detecting the more than one structural breaks.

$$\Delta co_t = \mu + \alpha co_{t-1} + \beta t + \theta_1 du1_t + \theta_2 du2_t + \sum_{j=1}^{k} dj\Delta co_{t-j} + \varepsilon_t \quad (8.9)$$

$$\Delta co_t = \mu + \alpha co_{t-1} + \beta t + \theta_1 du1_t + \theta_2 du2_t + \gamma_1 dt1_t + \sum_{j=1}^{k} dj\Delta co_{t-j} + \varepsilon_t \quad (8.10)$$

$$\Delta co_t = \mu + \alpha co_{t-1} + \beta t + \theta_1 du1_t + \theta_2 du2_t + \gamma_1 dt1_t + \gamma_2 dt2_t + \sum_{j=1}^{k} dj\Delta co_{t-j} + \varepsilon_t \quad (8.11)$$

On the same token, the mean shift is detected through du1 and du2, whereas the trend shift is detected through dt1 and dt2. Eq. (8.9) shows two breaks in intercept. Two breaks in intercept term and one break in trend are captured in Eq. (8.10). Finally, two breaks in intercept and two breaks in trend are captured in Eq. (8.11). The same grid search procedure is needed to implement the test for testing the unit root hypothesis. This is a procedure that starts from the estimating the basic equation through OLS and then test the unit root with null hypothesis $\alpha = 0$ for each potential structural break.

Clemente et al. (1998) Unit Root Test

Clemente et al. (1998) present another unit root test with two structural breaks. This is known as the Clemente et al. (1998) unit root test in time series literature. The null hypothesis of the test is that the series has unit root with breaks, whereas the alternative hypothesis is that the series is stationary or no unit root with breaks. The major advantage of this test is that we may apply it with unknown structural breaks. More clearly, a researcher does not require prior information about the structural breaks in the series. Clemente et al. (1998) argue that there may be two types of outlier in a data series: first, the outliers that may occur suddenly and second that may occur gradually in a time series. Therefore, these two different types of outliers should be treated differently. For this purpose, Clemente et al. (1998) propose two different models, AO model and IO model. The AO model will capture the sudden change, whereas the IO will capture the gradual change in the mean of the time series.

COINTEGRATION AND ERROR CORRECTION MODEL WITHOUT STRUCTURAL BREAKS

We have discussed in detail about the stationarity and nonstationarity in the last section. It is established in the time series literature that if we regress a nonstationary series on a stationary series, then we may get the spurious results despite the strong diagnostics of the regression. More clearly, we get spurious results even in the presence of high R^2, significant t-test and F-test. For example, if we regress the GDP of Bolivia on CO_2 emission of Pakistan, then we may get some good results despite the fact that there is no direct relation between GDP of Bolivia on the emission of Pakistan. This phenomenon is known as spurious regression. However, the regression will not be spurious if we can establish a long-run relationship, that is, two or more data series are cointegrated. Therefore, the understanding of cointegration is crucial in the context of time series. This section is an attempt in this way and will cover the issue of cointegration. Before moving on, it is important to mention that cointegration means the long-run relationship among the variables. It implies that if the nonstationary variables have a long-run relationship, then the results may not be spurious.

Tests of Cointegration

Once it is confirmed that the one or more underlying series are nonstationary then the next step is to test the cointegration. There are number of cointegration tests that are presented and used by the EKC studies in the literature of time series studies of EKC econometrics (see for example Farhani & Ozturk, 2015; Jalil & Feridun, 2011; Esteve and Tamarit 2012a; Esteve and Tamarit 2012b; Ozatac, Gokmenoglu, & Taspinar, 2017; Ozturk & Acaravci, 2013). The choice of test depends on the stationarity properties and structural breaks of time series data. Mainly, the tests of cointegration typically decided on the basis of unit root test.

Fig. 8.2 depicts the guidelines for the selection of the methodology of cointegration. As mentioned earlier, the unit root test is the natural start of a time series analysis. The unit root test may guide us about the stationary or nonstationary series. If the data series is stationary, then we call it integrated at level, that is, I(0). If the data series is nonstationary, then we may get a stationary series by taking the difference of the series. If we get a stationary at first difference, then we call it the integrated at first level, that is, I(1). The series that could be stationary at difference is called integrated at second order, that is, I(2).

If all data series in an underline analysis are I(0), then we apply a simple OLS or a linear regression like a conventional analysis. In this case, the short-run and long-run coefficients coincide. If data series are integrated at different level, that is, some are I(0) and some are I(1), then we will apply autoregressive distributed lag (ARDL) model. If all data series are integrated at the same level, that is, integrated at the first level, then we apply Engel-Granger cointegration test in a bivariate case and Johansen-Juselius cointegration test in the multivariate case. Toda and Yamamoto (1995) long-run causality could be estimated in the case of I(2) series. It is also important to note that if we cannot get the cointegration among the underlying series, then we shall estimate unrestricted VAR and estimate the Granger causality or impulse response function for the short-run analysis.

The following section will explain the procedure of cointegration both with structural breaks and without structural breaks. Specifically, we shall explain Engle-Granger cointegration procedure, Johansen-Juselius cointegration procedure and ARDL-bound testing in the stream of without structural breaks and Gregory and Hansen (1996) cointegration test along with Maki (2012) cointegration test in the stream of with structural breaks data series.

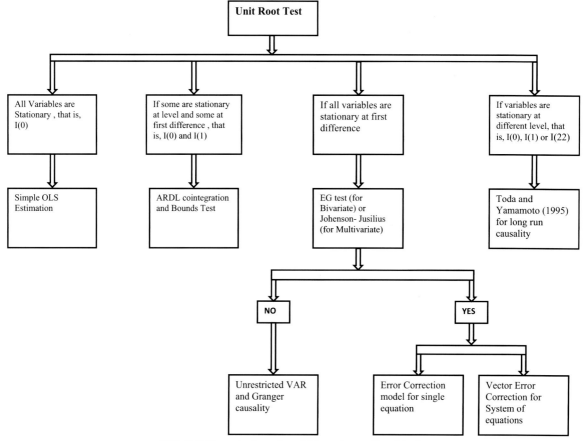

FIG. 8.2 The selection of model based on unit root test.

Engel-Granger cointegration test

The first test of cointegration was proposed by Engle and Granger (1987). This explicitly posts that how two nonstationary data series can be cointegrated. Let us suppose we estimate an EKC equation[2]

$$co_t = \beta_0 + \beta_1 \gamma_t + \beta_2 \gamma_t^2 + u_t \qquad (8.12)$$

The procedure of Engel-Granger cointegration test goes into two steps:
1. Estimate Eq. (8.12) through OLS, and obtain the residual series.
2. Test the stationarity property of residual series.

If the residual series is stationary at level, then we may conclude that the variables are cointegrated. More clearly, there exist a long-run relationship among

the variables of Eq. (8.12). Therefore, this regression will not be a spurious regression. However, generally, Engle-Granger cointegration test fails to capture more than one cointegration as given in Eq. (8.12). Most of the time, the researcher deals with multivariate models in the case of EKC. Therefore, it is not very useful in the context of EKC. Therefore, the most of the EKC literature concentrates on multivariate cointegration models, such as Johansen-Juselius procedure and ARDL cointegration test (for example Addullah et al., 2012; Lav et al., 2014; Onafowora & Owoyes, 2014; Ozatac et al., 2017; Ozturk & Acaravci, 2013).

Johansen cointegration test: Cointegration in multiple equations

As it is mentioned earlier, Engel-Granger cointegration test is not appropriate in a multivariate setup, and EKC remains a multivariate analysis in most of the cases. Generally, we take energy, trade, and other developments into account. Therefore, Johansen-Juselius

---

[2]It is a multivariate model. We explicit mentioned that the Engel-Granger test is not appropriate in the case of multivariate model. However, we take this equation for an example.

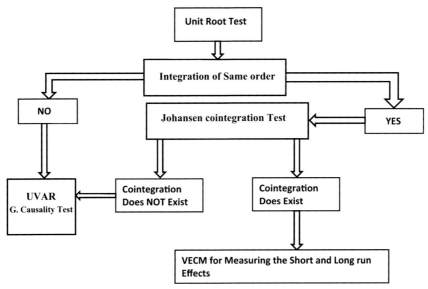

FIG. 8.3 Johansen-Juselius cointegration procedure.

cointegration test is the most appropriate technique in the case of multivariate analysis when the order of integration is same (Abdallah, Belloumi, & De Wolf, 2013; Abdou & Atya, 2013). There are several steps to follow to understand the Johansen-Juselius cointegration procedure and the estimation of error correction model. This particular section is spared for the understanding of this procedure.

Steps in Johansen cointegration. As mentioned earlier, most of economic data series are not stationary. Specially, the series that are considered to analyze the EKC, for example GDP, measure of environmental degradation, energy consumption, trade openness among others are almost nonstationary in almost all of the countries. It is well documented in the time series literature that the OLS regression will produce a spurious regression in the case of nonstationary series. In the case of nonstationary series, we try to establish a long-run relationship among the variables.

Step 1: Detecting the order of integration. The methodologies of detecting the long-run relationship depend on the order of integrations. For this purpose, we use ADF test, Philips-Perron test, or any other test depending on the situation. All series should be cointegrated of same order, ideally I(1), so that we may proceed further to apply Johansen-Juselius

cointegration test. However, if some of the series of I(0) or I(2), then we shall not apply Johansen-Juselius cointegration test. We let this discussion for a while. In nutshell, applying unit root test is the important step to start the procedure of cointegration testing (Fig. 8.3).

Step 2: Choosing the lag length. Second important step in this regard is the selection of appropriate lag length. It is important to mention here that almost all of the AR models are sensitive to the number of lag length. It implies the stability of the models crucially depend on the optimal lag lengths. It is established in the time series literature that the appropriate number of lags ensures the Gaussian error. We have already discussed that Gaussian error is free from the econometric issues such as heteroscedasticity, autocorrelation, and nonnormality. The major reason of these econometric issues due to non-Gaussian error is the omitted variable bias, which is very common in the case of EKC hypothesis testing in the empirical literature. Additionally, including and excluding the unnecessary lags also cause the omitted variable or redundant variable bias. Generally, the optimal lags are selected through a grid search procedure. The researchers estimate a VAR model of higher order and then reestimate the model by reducing the number of lags until zero lag. The optimal number of lags is decided on the basis of AIC or SBC.

The "Lags" in Economics.

It is well established that an independent variable rarely affect the dependent variable. There involves, in the most of case, a gap or lapse of time. This gap of time or lapse of time is called "lag." This is very true in the case of EKC hypothesis. Because, the turning points will not be achieved instantaneously. Therefore, the selection of lags is an integral part of the model selection.

SIGNIFICANCE OF LAGS

Three main reasons:
1. **Psychological·** A change in the independent variable does not necessarily lead to an immediate change in the dependent variable. The change in the independent variable may be transitory or permanent. The reaction to each change may be different depending on psychological reasons.
2. **Imperfect Knowledge**: Imperfect knowledge may lead to lags in a decision-making process, such as consumption and/or investment.
3. **Institutional:** Contractual obligations may prevent agents from changing their economic behavior.

Step 3: Appropriateness of the model.

The issue of entering the trend and intercept in the short-run and long-run equations should be discussed in estimating the model. Johansen (1995) discusses five possible ways. First, there is no intercept and time trend (deterministic trend) in cointegration equation. This model is applied when there are no deterministic components in the time series data. Obviously, practically, this is a very rare case. Second possibility postulates that there should be restricted intercept and there should not be a deterministic trend in a cointegration equation. Third is a common case, especially in the context of EKC. Johansen (1995) includes unrestricted intercept and no deterministic trend in cointegration equation and VAR model. Fourth one includes an unrestricted intercept and restricted trend in cointegration equation and VAR. Finally, an unrestricted intercept and quadratic trend are included in the cointegration equation in VAR model. Johansen (1995) notes that the first and fifth is a rare possibility to happen in the empirical literature of time series. Therefore, a researcher has to decide about the appropriateness of the model among the rest of three possibilities. According to Johansen (1995), a researcher relies on Pantula principle to select an appropriated model.

Step 4: Determining the number of cointegrating vector.

The next step is to determine the number of cointegrating vector or the number of long-run linear relationships. There is a possibility of "n-1" number of cointegrating vectors. Here n is the number of variables in the model. Johansen (1995) suggests a set of steps to determine the number of cointegrating vectors. Specifically, this approach calculates the long-run coefficients and the speed of adjustment. The speed of adjustment is generally known as the error correction term. The number of cointegrating vectors is determined through the set of two statistics. These statistics are called maximum Eigen values and the trace statistics.

Step 5: Estimating the error correction model.

Consider a bivariate model, for the simplicity, for CO_2 and GDP.

$$co_t = \beta_0 + \beta_1 y_t + \varepsilon_t \qquad (8.13)$$

We have to make sure that the both variables are integrated of order (1), that is, stationary at first level, test the existing of long-run relationship among two series with appropriate lags.

The cointegration equation between two variables is as follows:

$$\varepsilon_t = co_t - \beta_0 - \beta_1 y_t \qquad (8.14)$$

The next step is to derive the error correction model.

$$\Delta co_t = \beta co_t + \alpha co \varepsilon_t + \sum_{h=1}^{l} a_{1h} \Delta co_{t-h} + \sum_{h=1}^{l} b_{1h} \Delta y_{t-h} + v_{co_t} \qquad (8.15)$$

$$\Delta y_t - \beta y_t + \alpha y \varepsilon_t + \sum_{h=1}^{l} a_{2h} \Delta co_{t-h} + \sum_{h=1}^{l} b_{2h} \Delta y_{t-h} + v_{y_i} \qquad (8.16)$$

Here v_{cot} and v_{yt} are stationary white noise process. This analysis can further be extended to multivariate analysis.

Autoregressive distributed lag cointegration model

The discussion on cointegration is started with the concept of stationarity and nonstationarity. We have established that the time series must be cointegrated in the case of nonstationary series to avoid the problem of spurious regression. In this regard, we have discussed that how Engel-Granger cointegration test and Johansen-Juselius cointegration test can help us to avoid this issue. However, the issue is not as straightforward as we have discussed. There could be several possibilities, regarding the stationarity, in a multivariate model.

First, if all variables are stationary at level, that is I(0), then we shall run the regression through OLS methodology. In this case, the short-run coefficient and the long-run coefficients will be coincided. Second, if all variables are stationary at first difference, that is I(1), more clearly, they are integrated at same order but not cointegrated. Johansen cointegration procedure will be applied. Then we shall take the difference of the series and estimate the model through OLS to get the short-run coefficients. As we have mentioned earlier that if the series are not cointegrated, then we can find only the short-run elasticities.

Third, if there exist a long-run relationship, that is cointegration, then we shall proceed in two steps. We shall run an OLS model with level variables to estimate the long-run coefficients and then an error correction model to get the short-run coefficients.

Fourth, the complicated situation arises when some of the series are I(0) and some of data series are I(1). More clearly, the series are integrated of order 0, order 1, or fractionally cointegrated. When the situation is not clear about the level of integration, then Engel-Granger cointegration test and Johansen-Juselius cointegration test are not valid. In this situation, Pesaran, Shin, and Smith (2001) and Pesaran and Shin (1999) suggest to use ARDL or bound testing model for establishing a long-run relationship, cointegration, among the variables due to various advantages over the other methodologies of cointegration. For example, the researchers can use when there is a mix of I(0) and I(1) data series. This may be a plausible case if the researcher use the growth rate or change variables to measure the activities in the context of EKC. Furthermore, the interpretation of the outcome is really easily and straightforward as this estimation technique has one equation. Importantly, different lag structures can be assigned to different variables according to their optimality. Keeping the given advantages of ARDL estimation technique, a number of researchers used ARDL for examining the relationship between CO_2, income, and the square of income, that is, EKC (Acaravci & Ozturk, 2010; Ahmad et al., 2016; Dogan & Turkekul, 2016; Ertugrul, Cetin, Seker, & Dogan, 2016; Farhani & Ozturk, 2015; Jalil & Feridun, 2011; Lav et al., 2014; Onafowora & Owoyes, 2014; Ozatac et al., 2017; Ozturk & Acaravci, 2013 and many others).

Steps involved in estimating autoregressive distributed lag model. There are several steps involved in ARDL methodology. We shall narrate them briefly in the following lines.

Step 1: Testing the unit root. Although, ARDL can be applied in a mixture of I(0), I(1), or fractionally cointegrated data series. This test fails in the case of I(2) due to several reasons, which will be discussed in a while. Therefore, the first step is to test the unit root. That is why, as mentioned earlier, the unit root testing is the natural start of the time series analysis. If the data series is I(2) or higher, then we shall not use this methodology. The choice of the unit root test depends on the nature of the data series. If the data series is with structural breaks, then we shall use ADF test. If we are suspecting some structural breaks, then we shall move on Zivot and Andrew test or Clemente et al. (1998) unit root test.

Step 2: Specification of unrestricted error correction model. We formulate an unrestricted error correction model in the second step as follows:

$$\Delta co_t = \alpha_0 + \sum_{i=1}^{p} \alpha_i co_{t-i} + \sum_{j=1}^{q_1} \beta_j y_{1t-j} + \sum_{k=1}^{q_2} \gamma_k y_{t-k}^2 + \delta_0 co_{t-1} + \delta_1 y_{t-1} + \delta_2 y_{t-1}^2 + e_t$$

(8.17)

As we are not restricting the coefficients, Pesaran et al. (2001) termed this as unconditional or unrestricted error correction term (Fig. 8.4).

Step 3: Selection of appropriate lags. We have sufficiently discussed about the importance of lags in autoregressive models. There are a variety of criteria available in the empirical literature of time series. Specifically, the Akaike information criteria, Schwartz-Bayesian information criteria, and LR criteria are widely used. However, SBC is more popular among the researchers, specifically who has a short-span time series data, due to its parsimonious nature.

Step 4: The error must be serially independent. This step is a follow-up of the last step. More clearly, a Gaussian error is a requirement in ARDL methodology. Therefore, it is critical that the error should be serially independence. The serial independence may affect the choice of the number of lags; therefore, heteroscedasticity, autocorrelation, functional form, and the normality are tested to ensure the Gaussian error.

Step 5: Dynamic stability. As the autoregressive model is highly sensitive to number of lags, sample points, and number of variables, ensuring the stability of the dynamic model is critical. Brown, Durbin, and

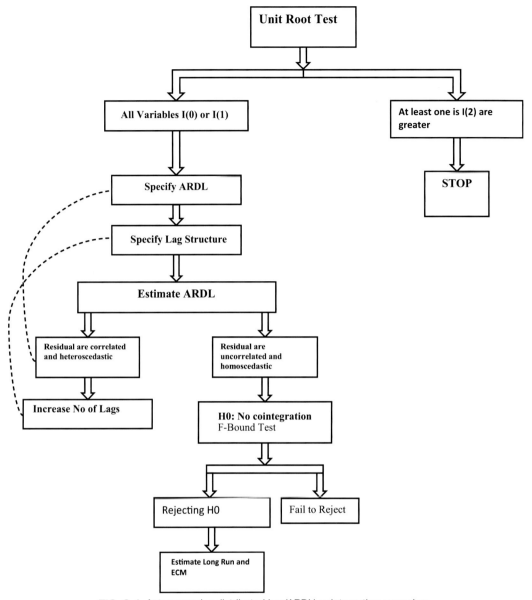

FIG. 8.4 Autoregressive distributed lag (ARDL) cointegration procedure.

Evans (1975) suggest the CUSUM and CUSUMSQ model for the stability of the model in the case of OLS. However, the detail of the ensuring the stability is beyond the scope, that is why we let it here.

Step 6: Performing the bound test/autoregressive distributed lag cointegration test. In this step, we shall shed light on performing the bound test or

cointegration test in the context of ARDL. Consider the Eq. (8.17) again

$$\Delta co_t = \alpha_0 + \sum_{i=1}^{p} \alpha_i co_{t-i} + \sum_{j=1}^{q_1} \beta_j y_{1t-j} + \sum_{k=1}^{q_2} \gamma_k y_{t-k}^2 + \delta_0 co_{t-1}$$
$$+ \delta_1 y_{t-1} + \delta_2 y_{t-1}^2 + e_t$$

$$(8.18)$$

Technically, we shall perform F-test for the null hypothesis $\delta_0 = \delta_1 = \delta_2 = 0$ against alternative hypothesis $\delta_0 \neq 0\ \delta_1 \neq 0$, and $\delta_2 \neq 0$. The null hypothesis is that there is no long-run relationship among the variables. It implies that if we fail to reject null hypothesis, then there will be no long relationship among the variables. Alternatively, there exists a long-run relationship among the variables. It important to mention here that we shall use the conventional F-statistic because the distribution is nonstandard and even asymptotic. Therefore, Pesaran et al. (2001) calculated the critical values for I(0) and I(1) at different levels of significance. If the calculated value is lower than I(0), then there will be no cointegration. On the other hand, there exists a long-run relationship if the calculated values fall above I(1). If the calculated values remain between I(0) and I(1), then there will no conclusion about the cointegration.

Step 7: Estimating the long-run model. If there exist a long-run relationship or the variables are cointegrated, then we shall proceed further to estimate the long-run model as well as the short-run dynamics along with the Error correction mechanism (ECM) term. The ECM, as mentioned earlier, shows the speed of adjustment of exogenous shocks of short runs toward the long-run equilibrium.

COINTEGRATION MODEL WITH STRUCTURAL BREAK

We have settled the case that conventional unit root tests and cointegration tests are not valid in the case of structural breaks or sudden change in the data series. The literature postulates that the power of cointegration test becomes weak in the presence of structural breaks in the data series (Campos et al. 1996; Gregory and Hansen 1996). There are two cointegration tests, which account the issue of structural breaks. These are Gregory and Hansen (1996) test and Maki (2012) test.

Gregory and Hansen Cointegration Test

This test is an extension of Engel-Granger cointegration test in the presence of structural breaks in the series. Gregory and Hansen (1996) present three cases in this regard as we have seen in the case of Zivot and Andrews (1992) unit root test. These cases are shift in level, shift in trend, and shift in regime. The null hypothesis in all the three cases is that there is no cointegration. It is important to note that Gregory-Hansen cointegration test is designed to capture the single structural break in the series. We specify EKC in the line of Gregory and Hansen (1996) to show the level shift.

$$CO_{2t} = \beta_0 + \beta_1 D_t + \alpha_1 y_t + \alpha_2 y_t^2 + u_t \tag{8.19}$$

Here D is dummy for the shift/level in the intercept at the time of structural break or sudden change.

Gregory and Hansen (1996) also present the second model to represent the shift in level with trend in the case of structural break. The shift in level with trend in the case of EKC could be specified as follows:

$$CO_{2t} = \beta_0 + \beta_1 D_t + \gamma t + \alpha_1 y_t + \alpha_2 y_t^2 + u_t \tag{8.20}$$

Here t with γ coefficient the time trend. The third Gregory-Hansen cointegration model shows the regime shift. The regime shift model in the case of EKC will look like as

$$CO_{2t} = \beta_0 + \beta_1 D_t + \gamma t + \alpha_1 y_t + \alpha_2 y_t^2 + \alpha_3 y_t D_t + u_t \tag{8.21}$$

α_3 is the coefficient of regime shift, that is, the shift in the slope of the EKC. The estimation will follow, as in the case of Zivot and Andrews (1992), the grid search procedure to calculate the test statistics for the testing of null hypothesis of no cointegration. Then these calculated values will be compared with the simulated values of Gregory-Hansen (1996) to decide about the rejection of null hypothesis. If we fail to reject the null hypothesis, then we shall move to Granger causality test in the case of structural breaks. If the null hypothesis is rejected, then we shall move to estimate the long-run estimates for EKC. In the case of cointegration, DOLS and FMOLS are the better options to calculate the short-run and long-run estimates (see Cho, Chu, & Yang, 2014; Ozcan, 2013).

Maki (2012) Cointegration Test

One of the major drawbacks of Gregory-Hansen cointegration test is that it takes only a single break. More clearly, the testing power of Gregory-Hansen cointegration test is better in the case of single break in the data series. However, this test does not perform well in the case of multiple breaks. To over the issue of multiple breaks, recently, Maki (2012) present a new cointegration test in the case of multiple unknown structural breaks. As mentioned earlier, the structural break can effect in three ways, that is, affect the level, trend, and slope. However, Maki (2012) propose our models in the case of multiple unknown breaks. These can be specifying, in the case of EKC, as follows:

$$CO_{2t} = \beta_0 + \sum_{i=1}^{p} \beta_i D_{i,t} + \alpha_1 y_t + \alpha_2 y_t^2 + u_t \tag{8.22}$$

$$CO_{2t} = \beta_0 + \sum_{i=1}^{p} \beta_i D_{i,t} + \gamma t + \alpha_1 y_t + \alpha_2 y_t^2 + u_t \tag{8.23}$$

$$CO_{2t} = \beta_0 + \sum_{i=1}^{p} \beta_i D_{i,t} + \alpha_1 \gamma_t + \alpha_2 \gamma_t^2 + \sum_{i=1}^{p} \gamma_t D_{i,t} + u_t \quad (8.24)$$

$$CO_{2t} = \beta_0 + \sum_{i=1}^{p} \beta_i D_{i,t} + \sum_{j=1}^{q} \gamma_j D_{j,t} + \alpha_1 \gamma_t + \alpha_2 \gamma_t^2$$
$$+ \sum_{i=1}^{p} \gamma_t D_{i,t} + u_t \quad (8.25)$$

Here β_i and γ_j are the representation of level shift and trend shift due to structural breaks in the time series, whereas α_i are the change in slope. D are the dummies for the structural breaks, whereas p and q are the maximum number of structural breaks.

The above four models capture all possibilities. For example, the first model captures the shift in the level due to the structural break, second captures the change in trend, third captures the change in slope, and fourth takes the account of every break in the data series. The test works under the null hypothesis of no cointegration against the alternative of cointegration among the underline variables. The test works the grid search procedure. More clearly, taking the example of first case, we estimate the first model to with maximum number of breaks. Then we obtain the residuals with 10% trimming. That is, we shall select the first structural break by minimizing the sum of square of the residuals. Then the suitable residual of suitable model will be tested, ADF test for test the null hypothesis of no cointegration. The simulated critical values are produced by Maki (2012).

CONCLUSION

The literature postulates that the time series analysis produces more information and target policy recommendations compared with the pooled or panel data analyses. Therefore, the time series analysis is more relevant in the case of policy-oriented research. Testing the EKC hypothesis is one of the examples in this case. More clearly, the panel data and the pooled data analyses provide generalized type of guidelines for the policy-makers. We cannot follow the generalized guideline in the case of environmental restoration. For example, in the case of panel data analysis, the policy guidelines for China, Pakistan, and Japan will be same. Surely, that should not be same. Therefore, the econometricians prefer to test the EKC hypothesis in country-specific cases for clear and country-specific guidance for the policy-makers. It is also important to note that there are number of researchers who are focusing to estimate the EKC in the country cases.

However, many of them are not well trained in the time series literature. Therefore, a wrong analysis may lead to a wrong policy implication. In this backdrop, this chapter focuses on to develop a comprehensive understanding of time series analysis.

Keeping this background in view, we highlight the general issues of time series analysis, which are related to test the EKC hypothesis. Specifically, we cover the issue of spurious regression, stationarity, unit root with and without structural breaks, cointegration with and without structural breaks. We have a feel that the researchers who are testing the EKC hypothesis in the country case should incorporate the newly developed methodologies, such as Gregory and Hansen cointegration test and Maki cointegration tests. We believe that these methodologies will make them to reestimate the EKC and recalculation of the turning points.

REFERENCES

Abdallah, K. B., Belloumi, M., & De Wolf, D. (2013). Indicators for sustainable energy development: A multivariate cointegration and causality analysis from Tunisian road transport sector. *Renewable and Sustainable Energy Reviews, 25,* 34−43.

Abdou, D. M. S., & Atya, E. M. (2013). Investigating the energy-environmental Kuznets curve: Evidence from Egypt. *International Journal of Green Economics, 7*(2), 103−115.

Acaravci, A., & Ozturk, I. (2010). On the relationship between energy consumption, CO_2 emissions and economic growth in Europe. *Energy, 35*(12), 5412−5420.

Ahmad, A., Zhao, Y., Shahbaz, M., Bano, S., Zhang, Z., Wang, S., et al. (2016). Carbon emissions, energy consumption and economic growth: An aggregate and disaggregate analysis of the Indian economy. *Energy Policy, 96,* 131−143.

Alshehry, A. S. (2015). Economic growth and environmental degradation in Saudi Arabia. *Journal of Economics and Sustainable Development, 6*(2), 33−44.

Box, G. E. P., & Jenkins, G. M. (1976). *Time series analysis: Forecasting and control* (Revised Edition). San Francisco: Holden Day.

Brown, R. L., Durbin, J., & Evans, J. M. (1975). Techniques for testing the constancy of regression relationships over time. *Journal of the Royal Statistical Society: Series B, 35,* 149−192.

Campos, J., Ericsson, N., & Hendry, D. (1996). Cointegration tests in the presence of structural breaks. *Journal of Econometrics, 70,* 187−220.

Chatfield, C. (2004). *The analysis of time series, an introduction.* Chapman & Hall.

Cho, C. H., Chu, Y. P., & Yang, H. Y. (2014). An environment Kuznets curve for GHG emissions: A panel cointegration analysis. *Energy Sources, Part B: Economics, Planning and Policy, 9*(2), 120−129.

Clemente, J., Montañés, A., & Reyes, M. (1998). Testing for a unit root in variables with a double change in the mean. *Economics Letters, 59,* 175−182.

Dickey, D. A., & Fuller, W. A. (1979). Distribution of the estimators for autoregressive time series with a unit root. *Journal of the American Statistical Association, 74*(366), 427−431.

Engle, R. (1982). Autoregressive conditional heteroscedasticity with estimates of the variance of United Kingdom inflation. *Econometrica, 50*(4), 987−1007.

Engle, R. F., & Granger, C. W. J. (1987). Co-integration and error correction: Representation, estimation, and testing. *Econometrica, 55,* 251−276.

Ertugrul, H. M., Cetin, M., Seker, F., & Dogan, E. (2016). The impact of trade openness on global carbon dioxide emissions: Evidence from the top ten emitters among developing countries. *Ecological Indicators, 67,* 543−555.

Esteve, V., & Tamarit, C. (2012a). Is there an environmental Kuznets curve for Spain? Fresh evidence from old data. *Economic Modelling, 29*(6), 2696−2703.

Esteve, V., & Tamarit, C. (2012b). Threshold cointegration and nonlinear adjustment between CO_2 and income: The environmental Kuznets curve in Spain, 1857−2007. *Energy Economics, 34*(6), 2148−2156.

Farhani, S., & Ozturk, I. (2015). Causal relationship between CO_2 emissions, real GDP, energy consumption, financial development, trade openness, and urbanization in Tunisia. *Environmental Science and Pollution Research, 22*(20), 15663−15676.

Gregory, A. W., & Hansen, B. E. (1996). Residual-based tests for cointegration in models with regime shifts. *Journal of Econometrics, 70,* 99−126.

Jalil, A., & Feridun, M. (2011). The impact of growth, energy and financial development on the environment in China: A cointegration analysis. *Energy Economics, 33*(2), 284−291.

Jalil, A., & Mahmud, S. (2009). Environment Kuznets curve for CO_2 emissions: A cointegration analysis for China. *Energy Policy, 37*(12), 5167−5172.

Johansen, S. (1995). *Likelihood-based inference in cointegrated vector autoregressive models.* New York: Oxford University Press.

Lau, L. S., Choong, C. K., & Eng, Y. K. (2014). Investigation of the environmental Kuznets curve for carbon emissions in Malaysia: Do foreign direct investment and trade matter? *Energy Policy, 68,* 490−497.

Lipford, J. W., & Yandle, B. (2010). Environmental Kuznets curves, carbon emissions, and public choice. *Environment and Development Economics, 15*(04), 417−438.

Lumsdaine, R. L., & Papell, D. H. (1997). Multiple trend breaks and the unit root hypothesis. *The Review of Economics and Statistics, 79,* 212−218.

Maki, D. (2012). Tests for cointegration allowing for an unknown number of breaks. *Economic Modelling, 29,* 2011−2015.

Onafowora, O. A., & Owoyes, O. (2014). Bounds testing approach to analysis of the environment Kuznets curve hypothesis. *Energy Economics, 44,* 47−62.

Ozatac, N., Gokmenoglu, K. K., & Taspinar, N. (2017). Testing the EKC hypothesis by considering trade openness, urbanization, and financial development: The case of Turkey. *Environmental Science and Pollution Research, 24*(20), 16690−16701.

Ozcan, B. (2013). The nexus between carbon emissions, energy consumption and economic growth in Middle East countries: A panel data analysis. *Energy Policy, 62,* 1138−1147.

Ozturk, I., & Acaravci, A. (2013). The long-run and causal analysis of energy, growth, openness and financial development on carbon emissions in Turkey. *Energy Economics, 36,* 262−267.

Ozturk, I., & Al-Mulali, U. (2015). Investigating the validity of the environmental Kuznets curve hypothesis in Cambodia. *Ecological Indicators, 57,* 324−330.

Pao, H. T., Yu, H. C., & Yang, Y. H. (2011). Modeling the CO_2 emissions, energy use, and economic growth in Russia. *Energy, 36*(8), 5094−5100.

Perron, P. (1989). The great crash, the oil price shock, and the unit root hypothesis. *Econometrica, 57*(6), 1361−1401.

Pesaran, M. H., & Shin, Y. (1999). An autoregressive distributed lag modelling approach to cointegration analysis. In S. Strom (Ed.), *Econometrics and economic theory in the 20th century: The Ragnar Frisch centennial symposium.* Cambridge: Cambridge University Press.

Pesaran, M. H., Shin, Y., & Smith, R. (2001). Bounds testing approaches to the analysis of level relationship. *Journal of Applied Econometrics, 16*(3), 174−189.

Phillips, P. C. B., & Perron, P. (1988). Testing for unit roots in time series regression. *Biometrika, 75,* 335−346.

Sims, C. (1980). Macroeconomics and reality. *Econometrica, 48*(1), 1−48.

Toda, H. Y., & Yamamoto, T. (1995). Statistical inference in vector autoregressions with possibly integrated processes. *Journal of Econometrics, 66*(1−2), 225−250.

Yavuz, N.Ç. (2014). CO_2 emission, energy consumption, and economic growth for Turkey: Evidence from a cointegration test with a structural break. *Energy Sources, Part B: Economics, Planning and Policy, 9*(3), 229−235.

Zivot, E., & Andrews, D. W. K. (1992). Further evidence on the great crash, the oil-price shock, and the unit-root hypothesis. *Journal of Business & Economic Statistics, 10,* 251−270.

FURTHER READING

Dong, B., Wang, F., & Guo, Y. (2016). The global EKCs. *International Review of Economics & Finance, 43,* 210−221.

Johansen, S. (1991). Estimation and hypothesis testing of cointegration vectors in Gaussian vector autoregressive models. *Econometrica, 59*(6), 1551−1580.

Johansen, S., & Juselius, K. (1990). Maximum likelihood estimation and inference on cointegration—with applications to the demand for money. *Oxford Bulletin of Economics & Statistics, 52*(2), 169−210.

CHAPTER 9

Panel Data Analysis (Stationarity, Cointegration, and Causality)

LIN-SEA LAU • CHEONG-FATT NG • SIEW-PONG CHEAH • CHEE-KEONG CHOONG

ESTIMATORS FOR PANEL DATA ANALYSIS APPLIED IN ENVIRONMENTAL KUZNETS CURVE

A number of literatures have been sparked in examining or evaluating the validity of environmental Kuznets curve (EKC) hypothesis by using various econometric techniques since the early 1980s. Over the past few decades, many different types of estimation techniques (cross section, time series, and/or panel data analysis) have been proposed in the econometric literature. For example, correlation test, regression analysis, unit root tests (Dickey-Fuller [DF], augmented Dickey-Fuller [ADF], Phillips-Perron [PP] tests, and so forth), cointegration and causality tests, and various tests for panel data models have been used.

The estimation techniques, nevertheless, are particularly focused on a few areas, namely the power of tests, stationarity, cointegration relationship, causality direction, cross-sectional dependence, homogeneity testing, and small sample bias problems. These issues are crucial and have attracted researchers' attention. The presence of any of the issues mentioned would substantially complicate estimation and inference in the analysis of EKC hypothesis.

This section of the chapter presents the development of econometric estimators for panel data analysis used in EKC.

Early Studies Using Pooled Ordinary Least Square and Generalized Least Square

Since the pioneering empirical work of Grossman and Krueger (1991) on EKC, a large number of studies have emerged applying various research methodologies. Most of the studies have used panel data to examine the existence of an inverted U-shaped relationship between pollution and economic growth for different groups of countries. Among these studies, many of the early research in the 1990s focused on the utilization of pooled ordinary least square (OLS) estimation in examining the validity of EKC hypothesis (e.g., Cropper & Griffiths, 1994; Panayotou, 1993; Shafik & Bandyopadhyay, 1992).

According to Stern, Common, and Barbier (1996), the estimation of EKC hypothesis involves two major problems. First, it is inappropriate to estimate the relationship between economic output and environmental degradation using a single equation model as there is an interdependence between the economy and the environment. In this respect, there is likely to be a bidirectional causality between the two variables. Using OLS to estimate the single-equation relationships with the existence of simultaneity problem produces biased and inconsistent results. Second, the use of poor-quality data may lead to the problems of heteroscedasticity. If the OLS estimation method is applied in the presence of heteroscedasticity problems, results obtained would be inefficient, although unbiased. In such a situation, diagnostic tests on the residuals are proposed to check for heteroscedasticity. In the meantime, the generalized least square (GLS) estimation instead of OLS should be used to achieve efficient outcomes.

Later, many EKC studies using panel data analysis such as Cole, Rayner, and Bates (1997), Selden and Song (1994), and Richmond and Kaufmann (2006) applied the GLS including the fixed effects and the random effects estimation, following Kmenta (1986). GLS first makes use of the estimated autocorrelation coefficient to transform observations to get rid of autocorrelation. It is then followed by dividing the transformed equation with the standard error of the residuals, to overcome the problem of heteroscedasticity.

In deciding to use whether fixed-effects or random-effects estimation, Hausman test can be used

Environmental Kuznets Curve (EKC). https://doi.org/10.1016/B978-0-12-816797-7.00009-6

(Hausman, 1978). The correlation between country effects and explanatory variables is an important issue that needs to be considered while choosing between the two estimations. If the correlation does not exist, the random-effects estimation is preferable. In contrast, in the presence of such correlation, the fixed-effects estimation would be more appropriate as there might be omitted variable bias.

Fully Modified Ordinary Least Square and Panel Dynamic Ordinary Least Square

In recent years, several modified OLS estimators have emerged to allow for (asymptotic standard) inference. These estimators include the fully modified ordinary least square (FMOLS) and the dynamic ordinary least square (DOLS), which have gained popularity among EKC researchers (e.g., Apergis & Payne, 2010; Farhani, Mrizak, Chaibi, & Rault, 2014; Jebli, Youssef, & Ozturk, 2016; Liu, Zhang, & Bae, 2017). Most importantly, both the FMOLS and DOLS estimators can overcome the problems of endogeneity in the regressors and the serial correlations in the error terms.

FMOLS is an estimator proposed by Phillips and Hansen (1990) with the aim of overcoming the problems created by the long-run correlation between the cointegrating equation and the stochastic regressor innovations. The standard FMOLS estimator for time series analysis was then extended into the pooled FMOLS by Phillips and Moon (1999). The modified dependent variable and the correction terms for serial correlation are shown below (Note that $\widehat{\Lambda}$ and $\widehat{\Omega}$ are long-run covariance).

$$\widetilde{y}_{it}^+ = \widetilde{y}_{it} - \widehat{\omega}_{it}\widehat{\Omega}_{22}^{-1}\widehat{u}_2 \qquad (9.1)$$

The estimated bias correction term is as follows:

$$\widehat{\lambda}_{12}^+ = \widehat{\lambda}_{12} - \widehat{\omega}_{12}\widehat{\Omega}_{22}^{-1}\widehat{\Lambda}_{22} \qquad (9.2)$$

The pooled FMOLS estimator is given by

$$\widehat{\beta}_{FP} = \left(\sum_{i=1}^{N}\sum_{t=1}^{T}\widetilde{X}_{it}X_{it}\right)^{-1}\sum_{i=1}^{N}\sum_{t=1}^{T}\left(\widetilde{X}_{it}y_{it}^+ - \chi_{12}^+\right) \qquad (9.3)$$

Specifically, DOLS is considered better than FMOLS because DOLS is able to remove the issues of endogeneity and autocorrelation using parametric approach but not nonparametric approach as in the case of FMOLS. The DOLS was first proposed by Saikkonen (1992) and Stock and Watson (1993). Kao and Chiang (2000) then extended the use of DOLS estimator to panel data set, which is coined as the pooled DOLS. Kao and Chiang (2000) used Monte Carlo simulations

to make a comparison among the OLS, FMOLS, and the DOLS estimators. They concluded that DOLS is a better estimator for both homogeneous and heterogeneous panels. In addition, the authors also highlighted the role of leads and lags in helping to reduce the bias of DOLS. Hence, it was concluded that DOLS estimator is more superior to the OLS and the FMOLS in estimating cointegrated panel regression. Kao and Chiang (2000) described the pooled DOLS estimator in which OLS is used to estimate an augmented cointegrating regression as follows:

$$y_{it} = \beta X_{it} + \sum_{j=-q_i}^{r_i}\Delta X_{it+j}\delta_i + \mu_{it} \qquad (9.4)$$

where y_{it} and X_{it} represent data removed from the individual deterministic trends. δ_i stands for the short-run dynamics coefficients, which are allowed to be cross-section specific. Let Z_{it} be the regressors formed by interacting the ΔX_{it+j} terms with cross-section dummy variables, and with $W_{it}' = \left(X_{it}'Z_{it}'\right)'$, the pooled DOLS estimator can be shown as below:

$$\begin{pmatrix}\beta_{DP} \\ \gamma_{DP}\end{pmatrix} = \left(\sum_{i=1}^{N}\sum_{t=1}^{T}W_{it}W_{it}'\right)^{-1}\left(\sum_{i=1}^{N}\sum_{t=1}^{T}W_{it}y_{it}'\right) \qquad (9.5)$$

Generalized Method of Moments

One of the most critical problems encountered in the econometric literature is pertaining to the estimation of linear regression models that contain heteroskedastic error of unknown functional form. In many of the time series and cross-sectional studies (e.g., Choi, 2001; Maddala & Wu, 1999), this issue has been widely discussed. Even though the form of the heteroscedasticity is unknown empirically, the ignorance of the issue in estimations (such as estimated generalized least squares[1]—EGLS) would cause inefficient estimators that result in erroneous inferences (Roy, 2002). Several researchers such as Robinson (1987) and Hidalgo (1992) suggested that this problem can be solved using nonparametric techniques. It is because such estimators are valid even with misspecified functional form. On the other hand, Rilstones (1991) proposed that Monte Carlo study can be used to make a comparison between the nonparametric EGLS estimators and the various parametric estimators using both correct and incorrect forms of heteroscedasticity.

Since the early 1990s, the issue of heteroscedasticity in the panel data estimations has been extensively

---

[1]The estimators of the technique, however, are asymptotically efficient (Rilstones, 1991).

discussed in the literature. Several studies examined the presence of heteroscedasticity in panel data analysis. These studies include Baltagi and Griffin (1988), Li and Stengos (1994), and Randolph (1988). Accordingly, Baltagi and Griffin (1988) examined the existence of heteroscedasticity through the individual specific error component by using parametric technique. However, Li and Stengos (1994) focused on the issue of heteroscedasticity in the unit-time-varying error component by using semiparametric method. It is concluded by both studies that the proposed EGLS estimators have the same asymptotic distribution as the true GLS estimator. On top of that, Li and Stengos (1994) argued that after conducting a Monte Carlo study, the finite sample properties of their estimator is found to be adequate as well. The results are inconsistent with the findings of Baltagi and Griffin (1988), in which their proposed procedure requires a large time component for the panel.

A semiparametric estimation procedure with unknown functional form in the individual specific errors was then proposed by Roy (2002). The newly recommended procedure does not need a large time component unlike the estimator suggested by Baltagi and Griffin (1988).[2] Three main findings were obtained. Firstly, efficiency is found in several standard estimators such as the proposed EGLS estimator (EGLS), the iterative EGLS estimator (EGLSB),[3] the standard GLS estimator for a one-way error components model (GLSH), the within or fixed effects estimator (WITHIN),[4] and the OLS estimator (OLS). Secondly, it is confirmed from the Monte Carlo study that the proposed estimator has an adequate relative efficiency. Nevertheless, it is sensitive to the selection of window-width. Thirdly, all of the estimators are found to behave in the similar pattern when it comes to size performance, that is, none of them overrejects or underrejects substantially.

Today, GMM panel data technique is applied in many EKC studies (e.g., Huang, Hwang, & Yang, 2008; Joshi & Beck, 2018; Khan, Zaman, & Zhang, 2016; Tamazian & Rao, 2010; Youssef, Hammoudeh, & Omri, 2016). This estimation technique was first proposed by Hansen (1982). Then, it was further improved by Arellano and Bond (1991), who introduced the difference GMM. A group of lagged explanatory variables are used as instruments for the corresponding variables in difference equation in the case of difference GMM. Later, Blundell and Bond (1998) claimed that small-sample and asymptotic properties of the difference estimator can be adversely affected by the issue of persistency in the explanatory variables. Thus, the difference estimator is combined with the original estimator to construct a system estimator, which is named as system GMM estimator.

The difference estimator is calculated using the moment conditions as follows:

$$E\big[y_{i,\,t-s}\big(\varepsilon_{i,\,t} - \varepsilon_{i,\,t-1}\big)\big] = 0 \text{ for } s \geq 2, \ t = 3, \ ..., T \quad (9.6)$$

$$E\big[X_{i,\,t-s}\big(\varepsilon_{i,\,t} - \varepsilon_{i,\,t-1}\big)\big] = 0 \text{ for } s \geq 2, \ t = 3, \ ...T \quad (9.7)$$

Two conditions must be fulfilled to use the lagged differences of the explanatory variables as instruments in levels equation. First, the error term is not serially correlated. Second, correlation does not exist between the difference in explanatory variables and the error terms despite the correlation between the levels of the explanatory variables and the country-specific error terms.

As a result, the following stationarity properties are produced:

$$E[y_{i,t+p}\eta_i] = E[y_{i,t+q}\eta_i] \text{ and } E[X_{i,t+p}\eta_i] = E[X_{i,t+q}\eta_i] \text{ for all } p \text{ and } q.$$

The additional moment conditions for the regression in levels are

$$E\big[\big(y_{i,\,t-s} - y_{i,\,t-s-1}\big)\big(\eta_i + \varepsilon_{i,\,t}\big)\big] = 0 \text{ for } s = 1 \quad (9.8)$$

$$E\big[\big(X_{i,\,t-s} - X_{i,\,t-s-1}\big)\big(\eta_i + \varepsilon_{i,\,t}\big)\big] = 0 \text{ for } s = 1 \quad (9.9)$$

Briefly, the system GMM estimator is obtained using the moment conditions in the above equations. According to Arellano and Bond (1991) and Blundell and Bond (1998), the validity of system GMM estimator can be checked using two tests. First, Sargan test can be carried out to test the validity of the instruments used. Second, the AR (2) test can be applied to check the existence of second-order autocorrelation.

GMM estimator has several advantages over other panel data estimators. First, Arellano and Bond (1991) confirm the fact that GMM estimator can optimally exploit all the linear moment restrictions that fulfill the assumption of no serial correlation in the errors. These moment restrictions that consist of individual effects, lagged dependent variables, and no strictly exogenous variables are vital in estimations. In addition, Hansen (1982) claimed that GMM estimator can provide consistency for models with nonlinear parameter.

---

[2]Roy (2002) has undertaken a Monte Carlo experiment to study the finite sample behavior of the proposed estimators.
[3]The model is proposed by Baltagi and Griffin (1988) with the assumption of homoscedastic error components.
[4]Refer to Baltagi (1985) for the detail of the technique.

Second, cross-sectional studies have two potential sources of bias, that is, the unobserved heterogeneity problem and the endogenous explanatory variables. By using both the cross-sectional and the time series variability, the GMM estimator can be seen as a promising alternative. For instance, the unobserved country-specific effects can be eliminated using GMM. In the meantime, it is also possible to correct for endogeneity problem in the first-differenced equations using a first-differenced GMM, proposed by Arellano and Bond (1991).

Third, GMM estimator can also overcome the problem of weak instrument. Blundell and Bond (1998) suggested that such a problem can lead to a large finite-sample bias while using the pooled cross-section regressions in estimating autoregressive models in the case of moderately persistent series from relatively short panels. Moreover, Blundell and Bond (1998) proved that by including more informative moment conditions that are valid under the reasonable stationarity restrictions on the initial condition process, the bias could be reduced greatly. Specifically, on top of the usual lagged levels as instruments for the equations in first-differences, the GMM estimator utilizes the lagged first-differences as instruments for the equations in levels.

Fourth, according to a number of studies such as Hsu and Liu (2006) and Mandariage and Poncet (2007, pp. 837−862), using OLS estimator with the presence of lagged dependent variables in equations would lead to the problem of inconsistency as the lagged dependent variables may be endogenous. These studies further propose that GMM estimator could eliminate the problems of heterogeneity and endogeneity. Most importantly, consistent and unbiased estimations could be produced eventually.

Pooled Mean Group Estimation

As mentioned by Pesaran and Smith (1995), the conventional estimation methods such as the fixed effects, the random effects, and the GMM have the purpose of correcting the fixed-effect heterogeneity issue that occur in the case of large N and small T panels. However, these estimators would produce inconsistent results as they do not take endogeneity caused by heterogeneity into consideration. To ensure consistent results, the mean group (MG) estimator introduced by Pesaran and Smith (1995) that tolerates differences in intercepts, slope, and error variances across groups can be used. However, the MG estimator does not take the issue of cross-sectional dependence into account. Alternatively, the pooled mean group (PMG) estimator

developed by Pesaran, Shin, and Smith (1999), which is more efficient due to the valid long-run restrictions, can be considered. Another advantage of the PMG over MG is that it is robust to the outliers and lag orders. Moreover, the PMG estimator requires long-run coefficients across cross sections to be similar. However, it allows for differences in the short-run coefficients, error variances, and the intercepts. Thus, the PMG estimator has become popular among the researchers studying the validity of the EKC hypothesis using panels in recent years (see, inter alia, Atasoy, 2017; Iwata, Okada, & Samreth, 2011; Li, Wang, & Zhao, 2016; Martínez Zarzoso & Bengochea-Morancho, 2004).

In the meantime, the PMG estimator is able to estimate a long-run relationship in a consistent and an efficient manner. It is unnecessary to examine the stationarity of the series as both stationary and nonstationary variables are allowed in the model with the condition that cointegration exists between variables (Pesaran & Shin, 1998). The model for the PMG estimator is presented as below:

$$y_{it} = \sum_{j=1}^{p} \gamma_{ij} y_{it-j} + \sum_{j=0}^{q} \beta_{ij} x_{it-j} + \mu_i + u_{it}$$

$$u_{it} = \rho_i' f_t + \varepsilon_{it} \qquad (9.10)$$

where x_{it} represents $k \times 1$ vector of explanatory variables for group i, μ_i is the fixed effects, β_{ij} denotes the coefficient vectors ($k \times 1$), γ_{ij} refers to coefficients of the lagged dependent variables, f_t is a vector of unobserved common shocks. Several important assumptions need to be made pertaining to the error term: (i) the error term in the second equation has a multifactor error structure; (ii) the errors are independently distributed across i and t; (iii) mean and variance of the errors are zero and larger than zero, respectively; and (iv) there is no correlation between the errors and the regressors or the unobserved common factors.

PANEL GRANGER CAUSALITY ANALYSIS

On top of the typical studies on the shape of the EKC, a strand of the literature dedicates to analyze the causality between pollution level and economic growth. Several causality testing approaches for panel data are available to analyze the nature of causality between pollution level (in particular, CO_2 emission) and economic growth, and, occasionally, along with other regressors.

Bivariate Panel Granger Causality Test

In a seminal paper, Granger (1969) developed a methodology for analyzing the causal relationships between time series. Suppose x_t and y_t are two stationary series. Then the following bivariate model:

$$y_t = \alpha + \sum_{k=1}^{K} \beta_k y_{t-k} + \sum_{k=1}^{K} \gamma_k x_{t-k} + \varepsilon_t \qquad (9.11)$$

can be used to test whether x causes y. The basic idea is that if past values of x are significant predictors of the current value of y even when past values of y have been included in the model, then x exerts a causal influence on y. Using model (9.11), one might easily test this causality based on an F-test with the following null hypothesis of noncausality:

$$H_0 : \gamma_1 = \ldots = \gamma_K = 0 \qquad (9.12)$$

If H_0 is rejected, one can conclude that causality runs from x to y. The x and y variables can of course be interchanged to test for causality in the other direction, and it is possible to observe bidirectional causality (or feedback relationship) between the time series.

With the advent of using panel data for econometric analysis, some authors attempted to extend the model (9.11) to its panel data counterpart. The pioneering work of Holtz-Eakin, Newey, and Rosen (1988) involved testing the hypothesis in Eq. (9.12) by using the following linear panel data model:

$$y_{i,t} = \alpha_i + \sum_{k=1}^{K} \beta_k y_{i,t-k} + \sum_{k=1}^{K} \gamma_k x_{i,t-k} + \varepsilon_{i,t} \qquad (9.13)$$

where i captures the individual specific effects across i. Now, the null hypothesis of noncausality $\gamma_1 = \ldots = \gamma_k = 0$ can be tested against the causality running from x to y for all the cross-sectional units. Notice that the coefficients β_k and γ_k in Eq. (9.13) are implicitly assumed to be fixed for all i. However, this approach may result in several problems. First, it is well known that the fixed effects estimator is biased and inconsistent in the dynamic panel data model when the data used is micropanel, for example, there are a large number of cross-sectional units observed over relatively short time periods (Nickell, 1981). Second, for panel data with finite time period, the Wald-type statistic with respect to Eq. (9.12) does not follow standard distribution (Hurlin & Venet, 2001). Finally, the panel Granger causality test proposed by Holtz-Eakin et al. (1988) imposes a homogeneous alternative hypothesis, which is a very strong hypothesis (Granger, 2003).

Given the limitations of the classic time series Granger procedure in the context of panel data, Coondoo and Dinda (2002) have ambitiously used Eq. (9.11) to a panel of 88 countries to detect the causality between income and emission. Because the application of Granger causality test is not common for testing EKC during the early 2000s, Coondoo and Dinda (2002) used various robustness measures, which include separating the 88 countries into 12 country groups and rely on the aggregate time series of each country group, and using the fixed-effect modeling approach to complement the OLS models.

To overcome the drawbacks of the panel Granger causality test proposed by Holtz-Eakin et al. (1988), Hurlin and Venet (2001), Hurlin (2004), and later Dumitrescu and Hurlin (2012) proposed testing the homogeneous noncausality (HNC) null hypothesis against the heterogeneous noncausality hypothesis (HENC) to complement the homogeneous causality (HC) hypothesis as in Holtz-Eakin et al. (1988). HENC allows some but not all of the individuals to Granger cause from x to y. The representation is now expressed as follows:

$$y_{i,t} = \alpha_i + \sum_{k=1}^{K} \beta_{ik} y_{i,t-k} + \sum_{k=1}^{K} \gamma_{ik} x_{i,t-k} + \varepsilon_{i,t} \qquad (9.14)$$

where β_{ik} and γ_{ik} are various coefficients of $y_{i,t-k}$ and $x_{i,t-k}$ for individual i, respectively. Coefficients are now allowed to differ across individuals but are assumed time-invariant. The lag order K is assumed to be identical for all individuals.

As in Granger (1969) and Holtz-Eakin et al. (1988), the procedure to determine the existence of causality is to test for significant effects of past values of x on the present value of y. The null hypothesis is therefore defined as:

$$H_0 : \gamma_{i1} = \ldots = \gamma_{iK} = 0 \qquad (9.15)$$

for $i = 1, \ldots, N$, which corresponds to the absence of causality for all individuals in the panel. The test assumes that there might be causality for some individuals but not necessarily for all. The alternative hypothesis is formulated as:

$$H_1 : \gamma_{i1} = \ldots = \gamma_{iK} = 0 \quad \forall i = 1, \ldots, N_1 \qquad (9.16)$$

$$\gamma_{i1} \neq 0 \ \text{ or } \ldots \text{ or } \ \gamma_{iK} \neq 0 \quad \forall i = N_1 + 1, \ldots, N$$

where $N_1 \in [0, N-1]$ is unknown. If $N_1 = 0$, there is causality for all individuals in the panel. N_1 is strictly smaller than N, otherwise there is no causality for all individuals, and H_1 reduces to H_0. Against this backdrop, Hurlin (2004) and Dumitrescu and Hurlin (2012) proposed the following procedure: Run the N individual regressions implicitly enclosed in Eq. (9.14), perform F-tests of the K linear hypotheses $\gamma_{i1} = \ldots = \gamma_{iK} = 0$ to retrieve W_i, and finally compute $\overline{W}$ as the average of the N individual Wald statistics:

$$\overline{W} = \frac{1}{N}\sum_{i=1}^{N}W_i \qquad (9.17)$$

where W_i is the standard adjusted Wald statistic for individual i observed during T period. Using Monte Carlo simulations, Dumitrescu and Hurlin (2012) showed that W is asymptotically well behaved and can genuinely be used to investigate panel causality.

Under the assumption that Wald statistics W_i are independently and identically distributed across individuals, it can be showed that the standardized statistic $\overline{Z}$ when $T \rightarrow \infty$ first and then $N \rightarrow \infty$ (sometimes interpreted as "T should be large relative to N") follows a standard normal distribution:

$$\overline{Z} = \sqrt{\frac{N}{2K}}\left(\overline{W} - K\right) \ \overrightarrow{T,N \rightarrow \infty} \ N(0,1) \qquad (9.18)$$

In addition, for a fixed T dimension with $T > 5 + 3K$, the approximated standardized statistic $\widetilde{Z}$ follows a standard normal distribution:

$$\widetilde{Z} = \sqrt{\frac{N}{2K}\cdot\frac{T-3K-5}{T-2K-3}}\left[\frac{T-3K-3}{T-3K-1}\cdot\overline{W} - K\right] \ \overrightarrow{N \rightarrow \infty} \ N(0,1) \qquad (9.19)$$

The testing procedure of the null hypothesis in Eqs. (9.15) and (9.16) is finally based on $\overline{Z}$ and $\widetilde{Z}$. If these are larger than the corresponding normal critical values, then one should reject H_0 and conclude that there is Granger causality. For large N and T panel data sets, $\overline{Z}$ can be reasonably considered. For large N but relatively small T data sets, $\widetilde{Z}$ should be favored. Using Monte Carlo simulations, Dumitrescu and Hurlin (2012) proved that the test exhibits very good finite sample properties. The new procedure by Hurlin (2004) and Dumitrescu and Hurlin (2012) also followed a standard Granger causality where the variables entering into the system need to be covariance-stationary.

Because of its features, the Dumitrescu-Hurlin procedure is commonly adopted by the studies searching for the growth-emission nexus in a bivariate setting. For instance, by using the DH procedure, Aye and Edoja (2017) found a unidirectional causality running from GDP growth to CO_2 emission in a panel of 31 developing countries over the period 1971−2013.

Error Correction Model−Based Framework and Granger Causality Test

As the questions of whether certain additional factors contribute to the traditional income-emission nexus

have gained importance from researchers and policy-makers, the application of Granger causality test in a multivariate setting has gained its popularity as well. These multivariate Granger causality analyses are generally based on the following linear equation:

$$Pollution_{it} = \beta_0 + \beta_1 Growth_{it} + \sum_{i=2}^{k}\beta_{it}X_{it} + \varepsilon_{it} \qquad (9.20)$$

where $Pollution$ normally indicates CO_2 emission, $Growth$ refers to economic growth, and X represents the additional variables that may contribute to the explanatory power of the equation. However, the least square estimate of Eq. (9.20) could be spurious as the variables are likely nonstationary at level, and there may exist cointegrating relationship among the variables. Likewise, the literature adopted techniques alike to the two-step procedure of Engle and Granger (1987) to estimate Eq. (9.20). The first step attempts to recover the lagged error correction term by estimating the cointegrating relation. There is a battery of estimation techniques that can be chosen such as the DOLS, FMOLS, the MG and the PMG estimators, GMM-type estimators, and so on. Among them, the FMOLS estimator is the most dominant choice by far. The second step then frames the error correction into a panel vector error correction model (VECM), which is the most commonly adopted technique among the causality testing analyses of EKC in a multivariate setting. The general representation of panel VECM is as follows:

$$\Delta Y_{it} = \delta_i d_t + \Pi_i ECT_{i,\,t-1} + \sum_{j=1}^{k-1}\theta_{ij}\Delta Y_{i,\,t-j} + \varepsilon_{it} \qquad (9.21)$$

where $Y_{it} = \left(y_{it1}, y_{it2}, ..., y_{itp}\right)'$ is a $p \times 1$ vector of endogenous variables of interest for individual i in period t, $\delta_i d_t$ is a $p \times 1$ vector representing the deterministic components of the model, and ε_{it} is the error term. As aforementioned, ECT is generated from Eq. (9.20) and the coefficient Π_i measures the speed of error adjustment of each i. A statistically significant Π_i with value between -1 and 0 indicates the existence of a long-run causality (strong exogeneity test). The detection of short-run causality can be conducted by computing the F-statistic on the coefficient matrix θ_{ij} (weak exogeneity test).

There are numerous applications of the ECM-based Granger causality framework in the analysis of EKC (see Antonakakis, Chatziantoniou, & Filis, 2017; Jaunky, 2011; Pao & Tsai, 2010; Saidi & Hammami, 2015). There are numerous studies apart from the ones mentioned above that have adopted various techniques to recover the ECM and ultimately to test for the panel causality. Interested readers can refer to some past applications of the DOLS (Farhani & Rejeb, 2012; Kais & Ben Mbarek, 2017), FMOLS (Ahmed, Rehman, & Ozturk, 2017; Farhani & Rejeb, 2012; Kais & Ben Mbarek, 2017; Shahbaz, Nasreen, Ahmed, & Hammoudeh, 2017), and the PMG (Lu, 2017) for panel causality analysis.

FIRST-GENERATION PANEL UNIT ROOT TESTS

This section introduces a battery of the first-generation panel unit root tests most commonly adopted in the literature of EKC, which include Levin, Lin, and Chu (2002), Im, Pesaran, and Shin (2003), and the Fisher-type tests.

Most of the panel unit root tests have evolved around the ADF (Dickey & Fuller, 1979) unit root test equation as follows:

$$\Delta y_{it} = \mu_i + \beta_i t + \rho_i y_{i,\,t-1} + \sum_{m=1}^{k_i} \alpha_{im} \Delta y_{i,\,t-m} + e_{it} \qquad (9.22)$$

where $\Delta y_{it} = y_{it} - y_{i,t-1}$, t is the time trend, k is the lag length, and e_{it} is the error term. By using Eq. (9.22) and assuming the dynamics of the autoregressive components are homogenous with cross-sectional independence for all panel units, Levin et al. (2002) proposed the following hypotheses:

$$H_0: \rho_i = 0$$
$$H_1: \rho_i < 0 \qquad (9.23)$$

If the null hypothesis $H_0: \rho = 0$ is not rejected, then the series is nonstationary at the corresponding order. Levin et al. (2002) proposed an alternative version by restricting ρ_i identical across all i as follows:

$$\Delta y_{it} = \mu_i + \beta_i t + \rho y_{i,\,t-1} + \sum_{m=1}^{k_i} \alpha_{im} \Delta y_{i,\,t-m} + e_{it} \qquad (9.24)$$

which assumes a common unit root across all individuals in the panel. The hypotheses are formulated as:

$$H_0: \rho_1 = \rho_2 = \ldots = \rho = 0$$
$$H_1: \rho_1 = \rho_2 = \ldots = \rho < 0 \qquad (9.25)$$

and the corresponding t-test statistic is computed as $t_\rho = \frac{\hat{\rho}}{SE(\hat{\rho})}$. The Monte Carlo exercise in Levin and Lin (1993) indicated that when there is no individual-specific fixed effect, the standard normal distribution

may provide a good approximation for the empirical distribution of the test statistic in relatively small samples. As Levin et al. (2002) noted, their panel-based unit root tests are more relevant for the panels that are in moderate size (i.e., $10 < N < 250$ and $25 < T < 250$).

Indeed, the alternative hypothesis H_1: $\rho_1 = \rho_2 = \ldots = \rho < 0$ is too strong to be held in the empirical studies in general (Maddala & Wu, 1999). Im et al. (2003) have overcome this limitation of the LLC test by computing the test statistic using MG approach. The resulting $\bar{Z}$ statistic is shown below:

$$\bar{Z} = \frac{\sqrt{N}[\bar{t} - E(\bar{t})]}{\sqrt{V(\bar{t})}}, \ \bar{t} = \frac{1}{N} \sum_{i=1}^{N} t_{\rho_i} \qquad (9.26)$$

where $E(\bar{t})$ and $V(\bar{t})$ are the mean and variance of each t_{ρ_i} statistic, respectively. Similar to LLC test, the Monte Carlo simulations in Im et al. (2003) showed that $\bar{Z}$ converges to a standard normal distribution and that the small sample performance of the IPS test is better than the LLC test. In addition, both LLC test and IPS test require $N/T \to 0$ for $N \to \infty$. If either N is small or if N is large relative to T, then the both tests show size distortions (Phillips & Moon, 1999).[5]

Finally, the Fisher-type test statistic (Choi, 2001; Maddala & Wu, 1999) combined P-values from unit root tests for each cross section i. The Fisher-type test statistic is as follows:

$$P = -2 \sum_{i=1}^{N} \ln p_i \qquad (9.27)$$

The test is asymptotically chi-square distributed with $2N$ degrees of freedom ($T_i \to \infty$ for finite N). A merit of the Fisher-type unit root tests is their handling of the unbalanced panels. Furthermore, the lag lengths of the ADF tests, PP tests or any other unit root tests are allowed to differ across individuals. However, a drawback of the test is that the P-values have to be obtained with Monte Carlo simulations.

Both the IPS and Fisher-type tests relaxed the restriction of the LLC test that i must be identical for all series under the alternative hypothesis. With respect to the size-adjusted power, the Fisher-type tests outperform the IPS test. It should be mentioned that in the presence

---

[5]The LLC test has high power if T is relatively large. This can be problematic, as one might infer stationarity for the whole panel, even if it is only true for a few individuals. On the other hand, it has low power for small T. In this case, one can conclude non-stationarity when in fact most series display stationary behavior.

of a linear time trend, the power of all tests decreases considerably (Hlouskova & Wagner, 2006).

As none of these panel unit root tests is devoid of statistical shortcomings in terms of size and power properties, it is very common that practitioners prefer to conduct several unit root tests and compare the outcomes. For example, Aye and Edoja (2017) applied the LLC, IPS, Fisher-ADF, and the Fisher-PP unit root tests in their study and conclude that all variables utilized are integrated of order one. Similarly, Pao and Tsai (2010) concluded that all variables in their study are integrated of order one according to the results of the LLC, IPS, and the Fisher-ADF tests. On the other hand, Antonakakis et al. (2017) chose to conduct the LLC and IPS tests, which confirm the existence of a unit root in the all variables used.

CROSS-SECTIONAL DEPENDENCY IN PANEL DATA

The EKC literature before the 2010s was relied on the first-generation unit root tests discussed above, which assume cross-sectional independence. Cross-sectional dependency (CSD) in panel data, especially macropanel has received a lot of attention in the recent years, especially after the 2010s. It is a type of correlation arising from common shocks with heterogeneous impacts across different countries, for instance the oil price shock in the 1970s and the global financial crisis in 2007. It also refers to the result of local spillover effects between regions or countries. Apart from this, spatial effects, omitted common effects or interactions within socioeconomic networks might be the reasons of cross-correlated errors (Atasoy, 2017). Eberhardt and Teal (2011) provided a detailed discussion for the issues within the cross-country empirics, whereas Moscone and Tosetti (2009) summarized the implications on the use of CSD tests.

Early literature on the EKC omits the testing for potential CSD problem in the panel data, especially before the 2010s, for example Dinda and Coondoo (2006), Richmond and Kaufmann (2006), and Apergis and Payne (2010). Ignoring CSD in estimation may lead to serious consequences, such as the loss of estimator efficiency and the invalid test statistics due to unaccounted dependency among the residuals. Recently, evidence from Apergis (2016), Atasoy (2017), and Apergis, Christou, and Gupta (2017) show that CSD often exists in panel data settings.

Testing for Cross-Sectional Dependency

A few tests are available for detecting CSD in the literature. The commonly applied test in the EKC literature is the Pesaran (2004) CD test (see Apergis, 2016; Apergis et al., 2017; Atasoy, 2017; Pablo-Romero, Sánchez-Braza, & Sánchez-Rivas, 2017; Pablo-Romero & Sánchez-Braza, 2017; Youssef et al., 2016). This test is based on a simple average of all pair-wise correlation coefficients of the OLS residuals obtained from the standard ADF regressions for each variable in the panel. The CD test statistic follows an asymptotically two-tailed standard normal distribution. The test examines the null hypothesis of cross-sectional independence or no CSD. The test statistic is calculated as follows:

$$CD = \sqrt{\frac{2T}{N(N-1)}} \left(\sum_{i=1}^{N-1} \sum_{j=i+1}^{N} \widehat{p}_{ij} \right) \sim N(0,1) \ i,j$$
$$= 1, 2, 3, ..., N \qquad (9.28)$$

where $\widehat{p}_{ij}$ is the sample estimate of the pair-wise correlation of the residuals obtained from the OLS.

By using the CD test, a number of EKC studies have shown that the data suffer from CSD (see Apergis, 2016; Apergis et al., 2017; Atasoy, 2017; Pablo-Romero et al., 2017; Pablo-Romero & Sánchez-Braza, 2017; Youssef et al., 2016).

Second-Generation Panel Unit Root Tests

As mentioned before, the CSD often exists in panel data analysis. Thus, the second-generation panel unit root test known as the cross-sectionally augmented IPS (CIPS) test by Pesaran (2007), which accounts for CSD is preferable over the first-generation panel unit root tests. The CIPS test is constructed from the results of the panel-member-specific ADF regressions, where cross-sectional averages of the dependent and independent variables are included in the model. Hence, the test is suitable to identify unit roots in heterogeneous panels. The test statistic has a nonstandard distribution with the null hypothesis of nonstationarity.

Specifically, the CIPS test uses the cross-section mean to proxy the common factor and constructs the test statistics based on the t-ratio of the OLS estimate of $b_i \left(\widehat{b}_i \right)$ in the following cross-sectional augmented Dickey-Fuller (CADF) regression:

$$D\gamma_{it} = \alpha_i + b_i \gamma_{i,\,t-1} + c_i \gamma_{t-1} + d_i D\gamma_t + \varepsilon_{it} \qquad (9.29)$$

A cross-sectional augmented version of the IPS test is based on the formula below:

$$CIPS(N,T) = t - bar = N^{-1}a_{i=1}^{N}t_i(N,T) \qquad (9.30)$$

where $t_i(N,T)$ is the ADF statistic across the cross section for the ith cross-section unit set by the t-ratio of b_i in Eq. (9.29).

There exist some recently published EKC studies (see, inter alia, Apergis, 2016; Apergis et al., 2017; Atasoy, 2017; Pablo-Romero et al., 2017; Pablo-Romero & Sánchez-Braza, 2017; Youssef et al., 2016), which adopted the CIPS test in the presence of CSD.

Second-Generation Panel Cointegration Tests

Similar to the panel unit root tests, the EKC literature before the 2010s was based on the first-generation panel cointegration tests, which assume cross-sectional independence while testing for the cointegration. There exist two commonly used panel cointegration tests proposed by Pedroni (1999, 2004) and Kao (1999). When CSD exists in the data, the results from the Pedroni and Kao cointegration tests may not be robust. Westerlund (2007) developed an error correction–based panel cointegration test that is robust even in the presence of the CSD. This is commonly known as the second-generation panel cointegration test. The underlying idea of the test is to examine the absence of cointegration by determining if error correction does exist among the individual panel members or among the whole panel.

Specifically, the test generates sample through the bootstrap method and uses a new sample to construct two-group mean statistics and two-panel statistics. This test searches for the presence of the error correction among the individual panel members or among the whole panel, based on the following model:

$$Dy_{it} = c_i + \alpha_i(y_{i,t-1} - b_ix_{it-1}) + \alpha_{j=1}^{P_i}\alpha_{ij}Dy_{it-j} + \alpha_{j=0}^{P_i}g_{ij}Dx_{it-j} + \varepsilon_{it} \qquad (9.31)$$

where α_i is the term of adjustment speed. The null hypothesis of the test assumes that there is no cointegrated relationship exists between the variables.

Robust Estimators in the Presence of Cross-Sectional Dependency

Recent EKC literature, using modern panel estimators, started addressing the potential problem caused by

CSD such as biased and inconsistent estimations results. For example, Atasoy (2017) uses two estimators robust to CSD, namely the Common Correlated Effects Mean Group (CCEMG) and the Augmented Mean Group (AMG) to examine the EKC hypothesis in the US panel. The CCEMG was introduced by Pesaran (2006), whereas the AMG was introduced by Bond and Eberhardt (2009) and Eberhardt and Teal (2010). These estimators are particularly for moderate time series and cross-section observations, where moderate typically means approximately 15 observations (Roodman, 2009). Both estimators are robust to CSD as they are concerned with the correlation across panel members. Other than CSD, the two estimators allow for heterogeneous slope coefficients, too.

The CCEMG estimator has the following model:

$$y_{it} = \alpha_{1i} + \beta_ix_{it} + \varphi_if_t + \varepsilon_{it} \qquad (9.32)$$

where y_{it} and x_{it} are observable variables, β_i is the country-specific slope on the observable regressor, α_{1i} is the group-fixed effects capturing time-invariant heterogeneity across groups, f_t is the unobserved common factor with heterogeneous factor loadings, which might be nonstationary and nonlinear to allow for cointegration, and ε_{it} is the error term.

Eq. (9.32) is then augmented with the cross-sectional averages of the dependent and independent variables as follows:

$$y_{it} = \alpha_{1i} + \beta_ix_{it} + \delta_i\bar{y}_{it} + \theta_i\bar{x}_{it} + \varphi_if_t + \varepsilon_{it} \qquad (9.33)$$

and estimated by using OLS for each cross section. The MG estimator for the CCE is obtained by averaging each coefficient over individual regression as follows:

$$CECMG = N^{-1}\sum_{i=1}^{N}\widehat{\beta}_i \qquad (9.34)$$

where $\widehat{\beta}_i$ is the estimates of coefficients in Eq. (9.33).

Turn to AMG, the main difference between CCEMG and AMG is the approximation method of the unobserved common factors, f_t is Eq. (9.32). The CCEMG estimator adopts the linear combinations of the cross-sectional averages of the unobserved common effects as well as the dependent and independent variables. Then, each individual coefficient is estimated by using OLS. AMG estimator uses a two-step method to estimate the unobserved common effects and to allow for cross-section dependence by incorporating the common dynamic effect parameter. The first step is to augment the equation with time dummies and make an estimation using first difference OLS as follows:

$$\Delta y_{it} = \alpha_{1i} + \beta_i \Delta x_{it} + \varphi_i f_t + \sum_{t=2}^{T} \pi_t DUMMY_t + \varepsilon_{it} \quad (9.35)$$

where Δ is the difference operator and π is the coefficient of the time dummies and referred as the common dynamic process. The next step is to augment the group-specific regression model with an explicit variable or a unit coefficient imposed on each group member. The imposition of a unit coefficient is implemented by subtracting the AMG estimator from the dependent variable.

The group-specific model parameters are averaged across the panel similar to those of the CCEMG estimator. Each regression includes an intercept, which captures the time-invariant fixed effects. The MG estimator for the AMG specified in Eq. (9.36) is similar to the MG estimator for CCEMG:

$$AMG = N^{-1} \sum_{i=1}^{N} \widetilde{\beta}_i \quad (9.36)$$

where $\widetilde{\beta}_i$ is the estimates of coefficients in Eq. (9.35).

CONCLUSION

Testing for the presence or validity of EKC hypothesis has been one of the most intensively studied topics. It is well documented that previous studies on the EKC hypothesis have reached mixed findings on the relationship between pollution and economic growth, depending on estimation techniques used. The inconsistent findings are likely due to econometric problems, which are not able to be fully captured or solved by a single methodology. For example, in the 1980s and 1990s, the studies have been trying to examine the presence of EKC hypothesis by using unit root and cointegration tests for a single country. Nevertheless, a number of studies have supported the use of panel data models, including panel least squares, fixed and random effect models, to examine the EKC hypothesis.

The use of panel data estimations rather than time series or cross-sectional estimations not only increases the total number of observations and their estimation power but also decreases the econometric problems arising from individual time series as well as static panel data techniques. After the development of various advanced methodologies, researchers have been able to provide justifications in explaining and predicting the EKC phenomenon. Hence, this study is aimed to

provide a comprehensive review of EKC literature by focusing on the use of various panel data techniques, their issues, problems, and advantages.

The study would suggest that researchers should be mindful when testing the EKC hypothesis using various estimation techniques. In particular, a rejection or acceptance of EKC hypothesis by a particular test does not imply an absolute conclusion. In fact, nonsupported EKC hypothesis may be due to some other reasons such as omitted variables and/or unobserved common factors. In addition, the validity of EKC findings should be complemented by using alternative techniques that are able to reduce econometric problems, such as heterogeneity and cross-section dependence.

REFERENCES

Ahmed, K., Rehman, M. U., & Ozturk, I. (2017). What drives carbon dioxide emissions in the long-run? Evidence from selected South Asian countries. *Renewable and Sustainable Energy Reviews, 70,* 1142−1153.

Antonakakis, N., Chatziantoniou, I., & Filis, G. (2017). Energy consumption, CO_2 emissions, and economic growth: An ethical dilemma. *Renewable and Sustainable Energy Reviews, 68,* 808−824. https://doi.org/10.1016/j.rser.2016.09.105.

Apergis, N. (2016). Environmental Kuznets curves: New evidence on both panel and country-level CO_2 emissions. *Energy Economics, 54,* 263−271. https://doi.org/10.1016/j.eneco.2015.12.007.

Apergis, N., Christou, C., & Gupta, R. (2017). Are there environmental Kuznets curves for US state-level CO_2 emissions? *Renewable and Sustainable Energy Reviews, 69,* 551−558. https://doi.org/10.1016/j.rser.2016.11.219.

Apergis, N., & Payne, J. E. (2010). Renewable energy consumption and economic growth: Evidence from a panel of OECD countries. *Energy Policy, 38*(1), 656−660.

Arellano, M., & Bond, S. (1991). Some tests of specification for panel data: Monte Carlo evidence and an application to employment equations. *The Review of Economic Studies, 58,* 277−297.

Atasoy, B. S. (2017). Testing the environmental Kuznets curve hypothesis across the US: Evidence from panel mean group estimators. *Renewable and Sustainable Energy Reviews, 77,* 731−747.

Aye, G. C., & Edoja, P. E. (2017). Effect of economic growth on CO_2 emission in developing countries: Evidence from a dynamic panel threshold model. *Cogent Economics & Finance, 5,* 1−22.

Baltagi, B. (1985). Pooling cross-sections with unequal time-series lengths. *Economics Letters, 18*(2−3), 133−136. https://doi.org/10.1016/0165-1765(85)90167-3.

Baltagi, B. H., & Griffin, M. A. (1988). Generalized error component model with heteroskedastic disturbances. *International Economic Review, 29,* 745−753.

Blundell, R., & Bond, S. (1998). Initial conditions and moment restrictions in dynamic panel data models. *Journal of Econometrics, 87*, 115−143.

Bond, S., & Eberhardt, M. (2009). Cross-section dependence in nonstationary panel models: A novel estimator. In *Paper presented at the Nordic econometrics conference in Lund*.

Choi, I. (2001). Unit root tests for panel data. *Journal of International Money and Finance, 20*, 249−272.

Cole, M. A., Rayner, A. J., & Bates, J. M. (1997). The environmental Kuznets curve: An empirical analysis. *Environment and Development Economics, 2*(4), 401−416.

Coondoo, D., & Dinda, S. (2002). Causality between income and emission: A country group specific econometric analysis. *Ecological Economics, 40*, 351−367.

Cropper, M., & Griffiths, C. (1994). The interaction of population growth and environmental quality. *The American Economic Review, 84*(2), 250−254.

Dickey, D. A., & Fuller, W. A. (1979). Distribution of the estimators for autoregressive time series with a unit root. *Journal of the American Statistical Association, 74*, 427−431.

Dinda, S., & Coondoo, D. (2006). Income and emission: A panel data-based cointegration analysis. *Ecological Economics, 57*(2), 167−181. https://doi.org/10.1016/j.ecolecon.2005.03.028.

Dumitrescu, E.-I., & Hurlin, C. (2012). Testing for Granger non-causality in heterogeneous panels. *Economic Modelling, 29*(4), 1450−1460.

Eberhardt, M., & Teal, F. (2010). *Productivity Analysis in Global Manufacturing Production*.

Eberhardt, M., & Teal, F. (2011). Econometrics for grumblers: A new look at the literature on cross-country growth empirics. *Journal of Economic Surveys, 25*(1), 109−155. https://doi.org/10.1111/j.1467-6419.2010.00624.x.

Engle, R. F., & Granger, C. W. J. (1987). Co-integration and error correction: representation, estimation, and testing. *Econometrica: journal of the Econometric Society*, 251−276.

Farhani, S., Mrizak, S., Chaibi, A., & Rault, C. (2014). The environmental Kuznets curve and sustainability: A panel data analysis. *Energy Policy, 71*, 189−198.

Farhani, S., & Rejeb, J. (2012). Link between economic growth and energy consumption in over 90 countries. *Interdisciplinary Journal of Contemporary Research in Business, 3*, 199−210.

Granger, C. W. J. (1969). Investigating causal relations by econometric models and crossspectral methods. *Econometrica, 37*(3), 424−438.

Granger, C. W. J. (2003). Some aspects of causal relationships. *Journal of Econometrics, 112*(1), 69−71.

Grossman, G. M., & Krueger, A. B. (1991). *Environmental impacts of a North American free trade agreement* (No. w3914). National Bureau of Economic Research.

Hansen, L. P. (1982). Large sample properties of generalized method of moments estimators. *Econometrica, 50*, 1029−1048.

Hausman, J. A. (1978). Specification tests in econometrics. *Econometrica: Journal of the Econometric Society*, 1251−1271.

Hidalgo, J. (1992). Adaptive estimation in time series regression models with heteroskedasticity of unknown form. *Econometric Theory, 8*, 161−187.

Hlouskova, J., & Wagner, M. (2006). The performance of panel unit root and stationarity tests: Results from a large scale simulation study. *Econometric Reviews, 25*, 85−116.

Holtz-Eakin, D., Newey, W., & Rosen, H. S. (1988). Estimating vector autoregressions with panel data. *Econometrica, 56*(6), 1371−1395.

Hsu, C. M., & Liu, W. C. (2006). The role of financial development in economic growth: The experience of Taiwan, Korea and Japan. *Journal of Asian Economics, 17*, 667−690.

Huang, B. N., Hwang, M. J., & Yang, C. W. (2008). Causal relationship between energy consumption and GDP growth revisited: A dynamic panel data approach. *Ecological Economics, 67*(1), 41−54.

Hurlin, C. (2004). *Testing granger causality in heterogeneous panel data models with fixed coefficients* (Vol. IX). Mimeo, University Paris.

Hurlin, C., & Venet, B. (2001). *Granger causality tests in panel data models with fixed coefficients* (Vol. IX). Mimeo, University Paris.

Im, S. K., Pesaran, M. H., & Shin, Y. (2003). Testing for unit roots in heterogeneous panels. *Journal of Econometrics, 115*(1), 53−74.

Iwata, H., Okada, K., & Samreth, S. (2011). A note on the environmental Kuznets curve for CO_2: A pooled mean group approach. *Applied Energy, 88*(5), 1986−1996.

Jaunky, V. C. (2011). The CO_2 emissions-income nexus: Evidence from rich countries. *Energy Policy, 39*, 1228−1240.

Jebli, M. B., Youssef, S. B., & Ozturk, I. (2016). Testing environmental Kuznets curve hypothesis: The role of renewable and non-renewable energy consumption and trade in OECD countries. *Ecological Indicators, 60*, 824−831.

Joshi, P., & Beck, K. (2018). Democracy and carbon dioxide emissions: Assessing the interactions of political and economic freedom and the environmental Kuznets curve. *Energy Research & Social Science, 39*, 46−54.

Kais, S., & Ben Mbarek, M. (2017). Dynamic relationship between CO_2 emissions, energy consumption and economic growth in three North African countries. *International Journal of Sustainable Energy, 36*, 840−854.

Kao, C. (1999). Spurious regression and residual-based tests for cointegration in panel data. *Journal of Econometrics, 90*(1), 1−44. https://doi.org/10.1016/s0304-4076(98)00023-2.

Kao, C., & Chiang, M. (2000). On the estimation and inference of a cointegrated regression in panel data. *Advances in Econometrics, 15*, 179−222.

Khan, S. A. R., Zaman, K., & Zhang, Y. (2016). The relationship between energy-resource depletion, climate change, health resources and the environmental Kuznets curve: Evidence from the panel of selected developed countries. *Renewable and Sustainable Energy Reviews, 62*, 468−477.

Kmenta, J. (1986). *Elements of econometrics*. London: Collier Macmillan.

Levin, A., & Lin, C. F. (1993). Unit root tests in panel data: new results. *University of California at San Diego, Economics Working Paper Series*.

Levin, A., Lin, C. F., & Chu, J. C. S. (2002). Unit root tests in panel data: Asymptotic and finite sample properties. *Journal of Econometrics, 108*(1), 1−24.

Li, Q., & Stengos, T. (1994). Adaptive estimation in the panel data error component model with heteroskedasticity of unknown form. *International Economic Review, 35*, 981−1000.

Liu, X., Zhang, S., & Bae, J. (2017). The impact of renewable energy and agriculture on carbon dioxide emissions: Investigating the environmental Kuznets curve in four selected ASEAN countries. *Journal of Cleaner Production, 164*, 1239−1247.

Li, T., Wang, Y., & Zhao, D. (2016). Environmental Kuznets curve in China: New evidence from dynamic panel analysis. *Energy Policy, 91*, 138−147.

Lu, W.-C. (2017). Greenhouse gas emissions, energy consumption and economic growth: A panel cointegration analysis for 16 Asian countries. *International Journal of Environmental Research and Public Health, 14*. https://doi.org/10.3390/ijerph14111436.

Maddala, G. S., & Wu, S. (1999). A comparative study of unit root tests with panel data and a new simple test. *Oxford Bulletin of Economics & Statistics, 61*, 631−652.

Mandariage, N., & Poncet, S. (2007). *FDI in Chinese cities: Spillovers and impact on growth*. The World Economy.

Martínez-Zarzoso, I., & Bengochea-Morancho, A. (2004). Pooled mean group estimation of an environmental Kuznets curve for CO_2. *Economics Letters, 82*(1), 121−126.

Moscone, F., & Tosetti, E. (2009). A review and comparison of tests of cross-section independence in panels. *Journal of Economic Surveys, 23*(3), 528−561. https://doi.org/10.1111/j.1467-6419.2008.00571.x.

Nickell, S. (1981). Biases in dynamic models with fixed effects. *Econometrica, 49*(6), 1417−1426.

Pablo-Romero, M., & Sánchez-Braza, A. (2017). Residential energy environmental Kuznets curve in the EU-28. *Energy, 125*, 44−54. https://doi.org/10.1016/j.energy.2017.02.091.

Pablo-Romero, M., Sánchez-Braza, A., & Sánchez-Rivas, J. (2017). Relationships between hotel and restaurant electricity consumption and tourism in 11 European Union countries. *Sustainability, 9*(11), 2109. https://doi.org/10.3390/su9112109.

Panayotou, T. (1993). *Empirical tests and policy analysis of environmental degradation at different stages of economic development* (No. 992927783402676). International Labour Organization.

Pao, H., & Tsai, C. (2010). CO_2 emissions, energy consumption and economic growth in BRIC countries. *Energy Policy, 38*(12), 7850−7860. https://doi.org/10.1016/j.enpol.2010.08.045.

Pedroni, P. (1999). Critical values for cointegration tests in heterogeneous panels with multiple regressors. *Oxford Bulletin of Economics & Statistics, 61*(s1), 653−670. https://doi.org/10.1111/1468-0084.0610s1653.

Pedroni, P. (2004). Panel cointegration: Asymptotic and finite sample properties of pooled time series tests with an application to the PPP hypothesis. *Econometric Theory, 20*(03). https://doi.org/10.1017/s0266466604203073.

Pesaran, M. H. (2004). *General diagnostic tests for cross-section dependence in panels*. https://doi.org/10.17863/CAM.5113.

Pesaran, M. (2006). Estimation and inference in large heterogeneous panels with a multifactor error structure. *Econometrica, 74*(4), 967−1012. https://doi.org/10.1111/j.1468-0262.2006.00692.x.

Pesaran, M. (2007). A simple panel unit root test in the presence of cross-section dependence. *Journal of Applied Econometrics, 22*(2), 265−312. https://doi.org/10.1002/jae.951.

Pesaran, M. H., & Shin, Y. (1998). An autoregressive distributed-lag modelling approach to cointegration analysis. *Econometric Society Monographs, 31*, 371−413.

Pesaran, M. H., Shin, Y., & Smith, R. P. (1999). Pooled mean group estimation of dynamic heterogeneous panels. *Journal of the American Statistical Association, 94*(446), 621−634.

Pesaran, M. H., & Smith, R. (1995). Estimating long-run relationships from dynamic heterogeneous panels. *Journal of Econometrics, 68*(1), 79−113.

Phillips, P. C., & Hansen, B. E. (1990). Statistical inference in instrumental variables regression with I (1) processes. *The Review of Economic Studies, 57*(1), 99−125.

Phillips, P. C., & Moon, H. R. (1999). Linear regression limit theory for nonstationary panel data. *Econometrica, 67*(5), 1057−1111.

Randolph, W. C. A. (1988). Transformation for heteroskedastic error components regression models. *Economics Letters, 27*, 349−354.

Richmond, A. K., & Kaufmann, R. K. (2006). Is there a turning point in the relationship between income and energy use and/or carbon emissions? *Ecological Economics, 56*(2), 176−189.

Rilstones, P. (1991). Some Monte Carlo evidence on the relative efficiency of parametric and semiparametric EGLS estimators. *Journal of Business & Economic Statistics, 9*, 179−187.

Robinson, P. M. (1987). Asymptotic efficient estimation in the presence of heteroskedasticity of unknown form. *Econometrica, 55*, 817−891.

Roodman, D. (2009). A note on the theme of too many instruments. *Oxford Bulletin of Economics & Statistics, 71*(1), 135−158. https://doi.org/10.1111/j.1468-0084.2008.00542.x.

Roy, N. (2002). Is adaptive estimation useful for panel models with heteroskedasticity in the individual specific error component? Some Monte Carlo evidence. *Econometric Reviews, 21*, 189−203.

Saidi, K., & Hammami, S. (2015). The impact of CO_2 emissions and economic growth on energy consumption in 58 countries. *Energy Reports, 1*, 62−70.

Saikkonen, P. (1992). Estimation and testing of cointegrated systems by an autoregressive approximation. *Econometric Theory, 8*(1), 1−27.

Selden, T. M., & Song, D. (1994). Environmental quality and development: Is there a Kuznets curve for air pollution emissions? *Journal of Environmental Economics and Management, 27*(2), 147−162.

Shafik, N., & Bandyopadhyay, S. (1992). *Economic growth and environmental quality: Time-series and cross-country evidence* (Vol. 904). World Bank Publications.

Shahbaz, M., Nasreen, S., Ahmed, K., & Hammoudeh, S. (2017). Trade openness–carbon emissions nexus: The importance of turning points of trade openness for country panels. *Energy Economics, 61*, 221–232.

Stern, D. I., Common, M. S., & Barbier, E. B. (1996). Economic growth and environmental degradation: The environmental Kuznets curve and sustainable development. *World Development, 24*(7), 1151–1160.

Stock, J. H., & Watson, M. W. (1993). A simple estimator of cointegrating vectors in higher order integrated systems. *Econometrica: Journal of the Econometric Society,* 783–820.

Tamazian, A., & Rao, B. B. (2010). Do economic, financial and institutional developments matter for environmental degradation? Evidence from transitional economies. *Energy Economics, 32*(1), 137–145.

Westerlund, J. (2007). Testing for error correction in panel data. *Oxford Bulletin of Economics & Statistics, 69*(6), 709–748. https://doi.org/10.1111/j.1468-0084.2007.00477.x.

Youssef, A. B., Hammoudeh, S., & Omri, A. (2016). Simultaneity modeling analysis of the environmental Kuznets curve hypothesis. *Energy Economics, 60*, 266–274.

CHAPTER 10

Most Up-to-Date Methodologic Approaches: Evidence from the Wavelet Coherence Approach

FAIK BILGILI, PHD • ERHAN MUĞALOĞLU, PHD • SEVDA KUŞKAYA, PHD •
H. HILAL BAĞLITAŞ, PHD • PELIN GENÇOĞLU, PHD

INTRODUCTION

The expected outcome of the 21st conference of parties (COP21) and 11th session of the parties to the Kyoto protocol (CMP11) held in Paris in December 2015 is a new international agreement on climate change targeting to keep the increase in global average temperatures below 2°C. Boden, Andres, and Marland (2017) report that the United States and China account for 15% and 30% global share of greenhouse gas emission (GHG), respectively. Besides, China, as the largest developing country, is responsible from 80% increase in worldwide CO_2 emissions backed with high economic growth rates since 2008 (Liu et al., 2013). This abrupt increase in Chinese GHG emissions, at the same time with a gradual emission reduction performance of the United States, proposes the relation between a country's economic growth and environmental position. The environmental Kuznets curve (EKC) hypothesis fundamentally and very basically stands for: While income per capita increases in a country, the environmental damage first rises and then diminishes (Stern, 2004). Panayotou (1993) first defines it as a Kuznets curve because it is similar to the seminal work of Kuznets (1955) explaining the relation between economic growth and inequality. The EKC relation considers an inverted U-shaped curve, which examines a mixed relationship between the environmental degradation and income.

An inverted U-shaped EKC hypothesis can be more generally defined as during a country's low-income level state, where the economy develops, first the environmental pollution increases and then starts to slow down till some income level threshold. After reaching this certain (threshold) income level, this turns out to be an environmental improvement (Grossman &

Krueger, 1995; Solarin, Al-Mulali, & Ozturk, 2017; Stern, 2004). The EKC hypothesis indicates an environmental evolutionary progress through a country's economic development that in early stages an economy performs more an energy and natural resource intensive growth where it emits more pollution, thus environmental quality diminishes. The increase in economic activity shifts the previous phase of production to industrial production, in which both technical progress and information technologies become important. At the last stage of economy, technical improvement triggers the new energy-efficient technologies and cleaner energy investments, which later overwhelm the negative effect of economic growth backed environmental degradation (Bo, 2011; Kaika & Zervas, 2013). As a result, environmental pollution growth would be gradually decreasing and later environmental quality would improve (Panayotou, 1993). These just defined three determining effects of the relationship between economic growth and environmental degradation: scale, structural, and technological effects (Bilgili, Koçak, & Bulut, 2016; Bilgili, Öztürk et al., 2016).

The aim of this chapter is to introduce the wavelet methodology, as a state of art method of causality testing, with an applied application. This wavelet-based application will be examining (1) (wavelet coherence) the interrelation between the cycles of income and CO_2 emission data series and (2) (partial wavelet coherence) reconsidering this relation by factoring out some other explanatory variables. Besides, (3) phase difference observations will be able to reveal a lead-lag analysis between these series. After presenting a brief review of the EKC hypothesis testing in Section 2, Section 3 covers the wavelet methodology and empirical data. Section 4 provides an empirical instance of

Environmental Kuznets Curve (EKC). https://doi.org/10.1016/B978-0-12-816797-7.00010-2

115

wavelet coherency and phase analysis with its results, and lastly Section 5 concludes.

LITERATURE REVIEW

In the early stages of EKC research literature, the validation test of an inverted U-shaped function of income per capita for different polluters has been examined by environmental economists including those of Panayotou (1993), Selden and Song (1994), Shafik (1994), Grossman and Krueger (1995), and Stern, Common, and Barbier (1996), and they all confirm an inverted U-shaped function for different pollution indicators. However, Stern et al. (1996) argue that first of all, due to the simultaneity (the choice of instrumental variables problem) and probable low data quality, the reduced form estimation of EKC parameters would not be econometrically strong enough for a future policy suggestion, and secondly, the EKC literature is mainly observed that it is valid for some specific pollution indicators rather than overall environmental degradation or quality. For instance, while one may expect an upward trend in sulfur dioxide (SO_2) emission, a flat deforestation rate in a specific time-interval is seen (Shafik, 1994). Selden and Song (1994) observed an exhibited EKC for air pollutants in the very long run; however, this may conclude very tricky policy implications, in favor of economic growth, if this very long run is beyond an irreversible environmental deterioration threshold (Panayotou, 1993).

Despite the fact that the EKC relationships are investigated for specific type of environmental damage indicators, such as deforestation, clean water, or ecologic footprint, and pollutants, such as SO_2, nitrous oxides (NO_2), and carbon dioxide (CO_2), Dinda (2004) reviews that inverted U-shaped curve of economic growth prevails mostly with respect to residential air pollutants, for example, carbon monoxide (CO) or solid particulate matter (SPM). Nahman and Antrobus (2005) and Acaravci and Akalin (2017, p. 10) provide brief review surveys of recent EKC studies that use different pollutants. Besides, Shahbaz and Sinha (2016) provide a detailed literature survey on EKC hypothesis testing, which set CO_2 emission as an independent variable. Shahbaz, Dube, Ozturk, and Jalil (2015), Shahbaz, Solarin, Sbia, and Bibi (2015) and Shahbaz and Sinha (2016) review that from the findings of both single- and multicountry data, there is no common convincing conclusion, which either validates or rejects the EKC relation for CO_2 emissions. On the other hand, Al-mulali, Weng-Wai, Sheau-Ting, and Mohammed (2015) investigate the existence of EKC hypothesis in 93 countries from various income levels by testing EKC hypothesis for ecologic footprint rather than air pollutants or CO_2 because carbon emissions are only responsible just for the half of total environmental degradation. This work has found out that there is an existing inverted U-shaped relationship except lower-middle income and low-income countries.

Shahbaz and Sinha (2016) state that although there are a vast number of studies for testing the validation of EKC hypothesis, most of these researches perform common model specifications. In addition to the specification of a model with respect to different environmental pollution indicators, we may also group the studies on the EKC validation regarding to (1) the estimation methodology used and depending on, (2) how income enters, and (3) which extra explanatory variables are inserted into estimation equation. The data methodology of EKC hypothesis can be primarily grouped into time series analysis and panel data analysis. Time series analysis typically stands for a single-country investigation of the EKC, whereas panel data analysis enables to simplify the result of EKC hypothesis for a group of countries. Each group of data reveals different results describing the association between environmental indicators and economic growth. The estimation methodology has a long variety; for example: ordinary least squares (OLS), autoregressive distributed lag (ARDL), dynamic OLS (DOLS), generalized method of moments (GMM), panel least squares (PLS), vector error correction method (VECM), etc. A recent summary of EKC validation literature using cointegration and causality tests is given in Table 10.1. The EKC hypothesis investigations of single- (time series) and multicountry (panel data) cases are presented in section (A) and (B), respectively. In addition to as usual methods, a summary of literature that uses methods for testing EKC hypothesis with parameter estimations other than causality and cointegration methods is presented in Table 10.2.

Income per capita, or sometimes rate of income growth (Coondoo & Dinda, 2002), can be included into EKC estimation equation with itself (linear), its second power (quadratic), and/or third power (cubic) forms. Depending on the estimated values of the parameters defining income, the EKC relation between environmental pollution indicators and income would be concluded. For instance, in any EKC testing equation, if all of the parameter estimates are not significantly different from zero, then there will be no relation between environmental degradation and economic growth. Moreover, there occurs an inverted U-shaped curve that confirms the EKC hypothesis, if the level and the second power of income have

TABLE 10.1
Environmental Kuznets Curve (EKC) Studies.

SECTION A: SINGLE-COUNTRY CASE (TIME SERIES)[a]

Author	Country	Data	Variable	Method	Result	EKC
Jalil and Feridun (2011)	China	1953–2006	FD, EG, EC, CE, RLL	ARDL	(EC, GDP, TO) $\Rightarrow$ CE	Valid
Nasir and Rehman (2011)	Pakistan	1972–2008	CE, GDP, TRD	Johansen cointegration	EG $\Rightarrow$ EC.	Valid
Xu (2011)	China	1990–2009	CE, GDP, INV, FIXC, M1	ARDL, Granger causality	(INV, GDP, FIXC) $\Rightarrow$ CE	Valid
Ahmed and Long (2012)	Pakistan	1971–2008	CE, EG, EC, TRL, POP	ARDL	(POP(−), TRL(+)) $\Rightarrow$ CE	Valid
Esteve and Tamarit (2012)	Spain	1857–2007	CE, GDP	Threshold cointegration and nonlinear adjustment	Nonlinear causality exists	Valid
Hamit-Haggar (2012)	Canada	1990–2007	EC, CE, EG	Granger causality	Short-run: CE $\Rightarrow$ EC; EG $\Rightarrow$ EC Long-run: EC $\Rightarrow$ EG.	Valid
Saboori, Sulaiman, and Mohd (2012)	Malaysia	1980–2009	GDP, CE, EG	ARDL, VECM	Long-run EG $\Rightarrow$ CE	Valid
Shahbaz, Lean, and Shabbir (2012)	Pakistan	1971–2009	EC, CE, EG, and TO	ARDL	EG $\Rightarrow$ CE	Valid
Abdallah, Belloumi, and Wolf (2013)	Tunisia	1980–2010	TVA, EC, CE	Johansen cointegration	TVA $\Rightarrow$ CE	Not valid
Saboori and Sulaiman (2013b)	Malaysia	1980–2009	ECD, EC, EG, CE	ARDL	EKC is valid for ECD but not for EC	Not observed
Shahbaz, Ozturk, Afza, and Ali (2013)	Turkey	1970–2010	EI, EG	VECM, Granger causality	EG/CE	Valid
Tiwari, Shahbaz, and Hye (2013)	India	1966–2011	EG, CE, TO	ARDL, Granger causality	(TO, EG) $\Leftrightarrow$ CE	Valid
Katircioğlu (2014)	Singapore	1971–2010	TOD, CE	Granger causality	TOD(−) $\Rightarrow$ CE.	Valid
Loganathan, Shahbaz, and Taha (2014)	Malaysia	1974–2010	CTAX, EG, EDEG	Granger causality	CTAX, EG, EDEG are cointegrated	Valid
Onafowora and Owoye (2014)	Selected countries	1970–2010	EG, EC, POP, TO, CE	ARDL, Granger causality	mixed results GDP ~ CE	Not observed
Shahbaz, Khraief, Uddin, and Ozturk (2014)	Bangladesh	1975–2010	ELC, CE, TO, IND	ARDL, IAA	ELC $\Rightarrow$ CE,IND; FD $\Rightarrow$ TO; TO $\Rightarrow$ IND	Valid
Shahbaz, Uddin, Rehman, and Imran (2014)	Tunisia	1971–2010	EG, EC, TO, CE	ARDL, IAA, VECM	EG $\Rightarrow$ CE; TO $\Rightarrow$ EC	Valid

Continued

TABLE 10.1
Environmental Kuznets Curve (EKC) Studies.—cont'd

Author	Country	Data	Variable	Method	Result	EKC
Tan, Lean, and Khan (2014)	Singapore	1975–2011	CE, EC, GDP	Granger causality	CE(−) ⇒ EG	Valid
Abid (2015)	Tunisia	1980–2009	EG, CE, GDP	Cointegrated VECM, Granger causality	EG ⇒ CE; GDP ⇔ EC	Not valid
Ahmed, Shahbaz, Qasim, and Long (2015)	Pakistan	1980–2013	EDEG, EG, EC, TO, POP	ARDL, VECM	(EG, TO) ⇒ EC; EG ⇔ EC	Valid
Ajmi, Hammoudeh, Nguyen, and Sato (2015)	G7 countries	1960–2010	CE, EC, GDP	Time-varying Granger causality	EKC is not supported	Not valid
Tang and Tan (2015)	Vietnam	1976–2009	CE, EC, FDI, EG	Cointegration, Granger causality	CE ⇔ GDP; EC ⇒ (CE, EC, FDI, GDP)	Valid
Ahmad et al. (2016)	India	1971–2014	EC, CE, EG	ARDL, VECM	EG ⇔ CE; EC(+) ⇒ CE Valid	
Bento and Moutinho (2016)	Italy	1960–2011	CE, GDP, NREP, REP, TRD	ARDL, Granger causality	TRD ⇒ (CE,NREP); GDP ⇒ NREP; NREP ⇒ REP	Valid
Ertugrul, Cetin, Seker, and Dogan (2016)	Selected developing countries	1971–2011	GDP, EC, TO, CE	ARDL with structural break, VECM, Granger causality	EKC is not valid for all selected countries	Not observed
Shahbaz, Solarin, and Ozturk (2016)	19 African countries	1971–2012	GDP, EI, GLB	ARDL	EKC is not valid for all countries	Not observed
Ahmad et al. (2017)	Croatia	1992–2011	GDP, CE	ARDL, VECM	In the short-run, EG ⇒ CE; in the long run EG ⇒ CE	Valid
Ali, Abdullah, and Azam (2017)	Malaysia	1971–2012	EC, CE, EG, FD, TO	ARDL, Granger causality	EC ⇔ CE; (EC, EG,FD) ⇒ CE	Valid
Charfeddine (2017)	Qatari	1970–2015	EG, EC, TO, UR, FD, EDEG, ECF	Cointegration with Markov switching model	EKC is valid only for CE	Not observed
Danish, Zhang, Wang, and Wang (2017)	Pakistan	1970–2012	REC, NREC, CE	ARDL, causality testing	CE ⇔ REC; CE ⇔ NREC	Valid
Kharbach and Chfadi (2017)	Morocco	2000–11	GDP, CE, POP, EG	Cointegration	EG(−) ⇒ CE	Valid

Study	Region/Country	Period	Variables	Method	Results	Validity
Shahbaz, Solarin, Hammoudeh, and Shahzad (2017)	USA	1960–2016	GDP, EX, IM, TO, BEC, CE	ARDL, VECM	Inverted U- and N-shaped EKC for EG and CE	Valid
Shahzad, Kumar, Zakaria, and Hurr (2017)	Pakistan	1971–2011	EC, TO, FD, CE	ARDL, Granger causality	EC ⇔ FD	Valid
Solarin et al. (2017)	China and India	1965–2013	CE, HEC;UR, GDP	ARDL, Granger causality	GDP ⇔ CE; HEC ⇔ CE	Valid
Bello, Solarin, and Yen (2018)	Malaysia	1971–2016	HEC, FFC, EDEG, GDP, UR	ARDL, VECM	GDP ⇔ EDEG	Valid
Zambrano-Monserrate, Silva-Zambrano, Davalos-Penafiel, Zambrano-Monserrate, and Ruano (2018)	Peru	1980–2011	GDP, CE, RELC, NGC PC	ARDL, VECM	(GDP, RELC, NGC, PC) ⇒ CE	Not valid

SECTION B: MULTICOUNTRY CASE (PANEL DATA)[a]

Study	Region/Country	Period	Variables	Method	Results	Validity
Pao and Tsai (2011)	BRICS countries	1980–2007	CE, EC, FDI, GDP, EG, FD	Panel cointegration	EC ⇒ CE; GDP ⇔ CE; GDP ⇔ EC;	Valid
Zilio and Recalde (2011)	Latin America and the Caribbean	1970–2007	EC, GDP	Cointegration approach	//	Not valid
Arouri, Youssef, Mhenni, and Rault (2012)	12 MENA countries	1981–2005	CE, EC, GDP	Panel cointegration	//	Not observed
Ozcan (2013)	12 Middle East countries	1990–2008	CE, GDP, EG, EC	Panel causality	Mixed results GDP ~ CE	Not observed
Saboori and Sulaiman (2013a)	ASEAN countries	1971–2009	CE, EC, GDP	ARDL, VECM	CE ⇔ CE	Valid
Farhani and Shahbaz (2014)	MENA region	1980–2009	RELC, NRELC, CE, GDP	Granger causality	(RELC, NRELC, GDP) ⇒ CE; EC ⇔ CE	Valid
Baek (2015)	12 major nuclear-generating countries	1980–2009	NE, EC, GDP, CE	Cointegration	NE(−) ⇒ CE	Valid
Omri, Daly, Rault, and Chaibi (2015)	MENA region	1990–2011	FD, CE, TRD, EG, TO	Simultaneous-equation panel	TO ⇔ FD; CE/FD; FD ⇒ EG; TO ⇒ CE	Valid

Continued

TABLE 10.1
Environmental Kuznets Curve (EKC) Studies.—cont'd

Author	Country	Data	Variable	Method	Result	EKC
Shahbaz, Dube et al. (2015), Shahbaz, Solarin et al. (2015)	Sub-Saharan African countries	1980–2012	EG, CE, EI, GDP	VECM, Granger causality	EI $\Rightarrow$ (EG,CE); EG $\Rightarrow$ CE	Valid
Al-Mulali and Ozturk (2016)	27 advanced economies	1990–2012	CE, GDP, REC, NREC, TO, UR, ENP	Panel Kao and Fisher cointegration, panel VECM, Granger causality	GDP, EC, UR $\Rightarrow$ CE; (REC, TO, ENP) (−) $\Rightarrow$ CE.	Valid
Al-Mulali, Ozturk, and Solarin (2016)	7 selected regions	1980–2010	CE, GDP, UR,TO, FD	Pedroni and Fisher type cointegration, Granger causality	EKC is restrictively valid	Valid
Apergis (2016)	15 countries	1960–2013	CE, GDP	Time varying panel cointegration	EKC is valid for only 12 countries	Valid
Dogan and Seker (2016)	Top renewable energy countries	1985–2011	GDP, REC, NREC, TO, FD, CE	Heterogeneous panel with cross-section dependence, cointegration	REC(−) $\Rightarrow$ CE; (NREC, GDP, FD) (+) $\Rightarrow$ CE	Valid
Jebli, Youssef, and Ozturk (2016)	OECD countries	1980–2010	CE, GDP, REC, NREC, IM, EX	Panel Granger causality	REC $\Leftrightarrow$ IM; REC $\Leftrightarrow$ NREC; NREC $\Leftrightarrow$ TRD; EX $\Rightarrow$ REC; TRD $\Rightarrow$ CE; GDP $\Rightarrow$ REC.	Valid
Moutinho and Robaina (2016)	European countries	1991–2010	REP, CE, GDP	Cointegration and causality	EKC is mainly determined by REP between 2001 and 2010	Valid
Youssef, Hammoudeh, and Omri (2016)	Countries within income levels	1990–2012	EDEG, GDP, EG, EC, FD	Simultaneous two-equation models and causality	EG $\Leftrightarrow$ EDEG	Valid
Ahmed (2017)	BRICS countries	1991–2013	EC, TO, EG, FD	Panel cointegration and causality	EKC is valid between (EC,FD); (EC,TO)	Valid
Ahmed, Rehman, and Ozturk (2017)	Selected South Asian countries	1971–2013	CE, EC, GDP, TO, POP	Panel cointegration	EC, TO, POP $\Rightarrow$ CE; EC $\Leftrightarrow$ TO	Valid

Study	Countries	Period	Variables	Methodology	Relationship	Validity
Charfeddine and Mrabet (2017)	19 MENA countries	1975–2007	ECF, EDEG, GDP	Granger causality, VECM	ECF ⇔ GDP	Valid
Dong, Sun, and Hochman (2017)	BRICS countries	1985–2016	CE, GDP, NGC, REC	Cointegration, AMG	CE ⇔ (NGC, REC)	Valid
Zhang et al. (2017)	China	1990–2014	EG, WPD	Granger causality, cointegration	EG ⇔ WPD	Valid
Zoundi (2017)	25 selected African countries	1980–2012	CE, GDP, RELC, POP, EC	ARDL, causality	GDP(+) ⇒ CE; in the long-run RELC(−) ⇒ CE	Not valid
Boutabba, Diaw, and Lessoua (2018)	17 sub-Saharan African countries	1995–2013	IGT, CE, GDP, CE	Panel cointegration, causality	In the long-run (GDP, IGT) ⇔ CE; in the short-run IGT ⇒ CE	Valid
Le and Quah (2018)	14 countries selected from Asia and Pacific	1984–2012	CE, EC, EG	Cointegration, Granger causality	EKC is valid only for developed countries	Not observed
Saleem, Jiandong, Zaman, Elashkar, and Shoukry (2018)	Next 11 countries	1975–2015	ED, POP, EDEG, GDP, EG	Cointegration	EDEG ⇔ GDP; ED ⇒ EG	Valid
Ulucak and Bilgili (2018)	Countries within income levels	1961–2013	GDP, ECP	Cointegration	Cointegrated relation between GDP and EFP	Valid

[a] See Notes (1, 2, 3) given in Appendix.

TABLE 10.2
A Summary of Environmental Kuznets Curve (EKC) Validation Literature (With Methods Other Than Causality Methods).[a]

Author(s)	Country	Period	Variables	Methodology	Conclusion
Grossman and Krueger (1991)	NAFTA	1972–88	Economic growth and air quality	Panel data	N-shaped EKC (for SO_2, and SPM dark matter)
Shafik and Bandyopadhyay (1992)	149 countries	1960–90	Economic growth and environmental quality	Panel data	Positive relation
Panayotou (1993)	30 developed and developing countries	1982–94	Economic development and environmental degradation	Cross-section data	Inverted U-shaped EKC
Grossman and Krueger (1995)	42 countries	1977–88	Income and environmental indicators	Regression	N-shaped EKC (for relationship with arsenic and income)
Holtz-Eakin and Selden (1995)	130 countries	1951–86	Economic growth and CO_2	Panel data	Positive relation
Roberts and Grimes (1997)	47 countries	1962–91	The economic development and CO_2	OLS	Inverted U-shaped EKC (for high income countries)
De Bruyn, Bergh, and Opschoor (1998)	Netherlands, UK, US, and Western Germany	1961–93	Economic growth and emissions	Panel data	Inverted U-shaped EKC
Kaufman et al. (1998)	23 countries	1974–89	Income and SO_2	Panel data, pooled OLS	Inverted U-shaped EKC
Suri and Chapman (1998)	33 counties	1971–91	Economic growth, trade, and pollutants	FGLS	Inverted U-shaped EKC
Torras and Boyce (1998)	19–42 countries	1977–91	Income, inequality, and pollution	Regression	Inverted U-shaped EKC
Galeotti and Lanza (1999)	110 countries	1971–96	GDP per capita and CO_2	Panel data	Inverted U-shaped EKC
Koop and Tole (1999)	76 developing countries	1961–92	GDP per capita and deforestation	Regression	Nonsignificant relationship
List and Gallet (1999)	48 US states	1929–94	Income and emission	Panel data	Inverted U-shaped EKC
Dinda, Coondoo, and Pal (2000)	33 countries	1979–82, 1983 –86, 1987–90	Economic growth and air quality	OLS and LAE regressions	U-shaped EKC
Hettige, Mani, and Wheeler (2000)	13 countries	1975–94	Economic development and industrial pollution	OLS	Inverted U-shaped EKC (without water pollutant)
Panayotou, Peterson, and Sachs (2000)	17 OECD countries	1870–1994	Income and CO_2	FGLS panel data	Inverted U-shaped EKC
Dijkgraaf and Vollebergh (2001)	24 OECD countries	1960–97	GDP per capita and pollution	Panel data	Inverted U-shaped EKC

Study	Countries	Period	Variables	Method	Result
Seppala, Haukioja, and Paivo-oja (2001)	Germany, Japan, US, Netherlands, Finland	1975–94	Economic growth and direct material flows	Regression	Nonsignificant relationship
Stern and Common (2001)	73 countries	1960–90	GDP per capita and SO_2	OLS, GLS	Inverted U-shaped EKC
Harbaugh, Levinson, and Wilson (2002)	45 countries	1971–92	National income and pollution	Regression	Inverted U-shaped EKC
Lindmark (2002)	Sweden	1870–1997	Economic growth, technology, oil price, and CO_2	Structural time series	Inverted U-shaped EKC
Halkos (2003)	73 OECD and non-OECD countries	1960–90	Economic growth and SO_2	GMM	Inverted U-shaped EKC
Bhattarai and Hammig (2004)	20 Latin American, 12 Asian, and 23 African countries	1980–95	National income and deforestation	Panel data	Inverted U-shaped EKC
Cole (2004)	18 OECD countries	1980–97	GDP and air pollutant, trade	Regression	Inverted U-shaped EKC
Martínez-Zarzoso and Bengochea-Morancho (2004)	22 OECD countries	1975–98	Income and CO_2	Panel data	N-shaped EKC
Shi (2004)	50 countries	1951–99	GNI and CO_2	Panel	Inverted U-shaped EKC
Aldy (2005)	US states	1960–99	Income and CO_2	OLS and FGLS	Inverted U-shaped EKC
Bertinelli and Strobl (2005)	122 countries	1950–90	GDP per capita, CO_2, and SO_2	Semiparametric kernel regression	U-shaped EKC
Galeotti and Lanza (2005)	108 countries	1971–95	GDP per capita and CO_2 per capita	Linear and log-linear model	N-shaped EKC
Azomahou, Laisney, and Van (2006)	100 countries	1960–96	GDP per capita and CO_2	Panel (nonparametric)	Positive relation
Culas (2007)	14 tropical developing countries from Latin America, Africa, and Asia	1972–74	Income and deforestation	Pooled regression	Inverted U-shaped EKC (only Latin America)
Kunnas and Myllyntaus (2007)	Finland	1800–2003	Economic growth and air pollution	Regression	Positive relation
Managi and Jena (2008)	India	1991–2003	Income and environmental productivity	Panel	Negative income effect
Akbostancı, Türüt-Aşık, and Tunç (2009)	Turkey	1992–2001	Income and environmental degradation	Panel	N-shaped EKC
Aslanidis and Iranzo (2009)	77 non-OECD countries	1971–97	Income and environmental degradation	Regression	Positive relation (for low-income countries)

Continued

TABLE 10.2
A Summary of Environmental Kuznets Curve (EKC) Validation Literature (With Methods Other Than Causality Methods).[a]—cont'd

Author(s)	Country	Period	Variables	Methodology	Conclusion
Lee, Chiu, and Sun (2009)	89 countries	1960–2000	Income and environmental quality	GMM	N-shaped EKC (for whole) Inverted U-shaped EKC (for middle income, American European countries)
He and Richard (2010)	Canada	1948–2004	GDP per capita and CO_2 per capita	Semiparametric and flexible nonlinear parametric model	Positive relation
Bo, Suying, Junbiao, and Haishan (2011)	China	2000–08	Agricultural output per capita and carbon intensity of agricultural land	Regression	Inverted U-shaped EKC
Gao (2011)	Hennan providence (China)	1994–2009	GDP per capita and pollution emissions	Regression	Inverted U-shaped EKC (only the industrial waste water emission)
Iwata, Okada, and Samreth (2011)	28 countries	1960–2003	GDP per capita, CO_2, and nuclear	Pooled mean group	Positive relation
Borhan and Ahmed (2012)	Malaysia	1996–2006	GDP per capita and water pollution	Simultaneous equations	EKC valid
Borhan, Ahmed, and Hitam (2012)	ASEAN-8	1965–2010	GDP per capita and CO_2	Simultaneous equations	EKC valid
Hitam and Borhan (2012)	Malaysia	1965–2010	GDP growth, FDI, and the environmental degradation	Regression	EKC valid
Apergis and Ozturk (2015)	14 Asian countries	1990–2011	GDP per capita and CO_2	GMM	Inverted U-shaped EKC
Sahli and Rejeb (2015)	21 MENA countries	1996–2013	GDP per capita and CO_2	Dynamic panel data	EKC valid
Ersin (2016)	13 developed countries	1870–2011	GDP per capita and CO_2	Dynamic panel-STAR	EKC valid
Ozokcu and Özdemir (2017)	26 OECD countries	1980–2010	Economic growth, energy, and environmental degradation	Panel data	N-shaped EKC and inverted- N-shaped EKC
Adu and Denkyirah (2018)	7 West African countries	1970–2013	Economic growth and environmental pollution	Panel data	Nonsignificant relationship

[a] See Notes (4) presented in Appendix.

significantly positive and negative parameter estimates, respectively. Empirical applications, which analyze the confirmation of EKC hypothesis in China, reveal consistent results with the inclusion of quadratic and cubic form of income. The EKC research papers, which use the quadratic form of income in estimation equation, mostly support an inverted U-shaped function of income particularly for the post-1970 period CO_2 emissions of China, including those of Jalil and Mahmud (2009), Pao and Tsai (2011), Jalil and Feridun (2011), Jayanthakumaran, Verma, and Liu (2012), Onafowora and Owoye (2014), Ertugrul et al. (2016), and Jin, Duan, Shi, and Ju (2017).

Finally, the EKC estimation equations usually consist of other explanatory variables other than income per capita. These most frequently included independent variables such as trade openness, energy consumption, renewable energy consumption and/or fossil fuel consumption, urbanization, and sectoral shares in national income. Suri and Chapman (1998) notify that the EKC relation exposes by the trade flow of manufacturing goods from industrialized to developing countries and vice versa. Energy consumption especially based on fossil fuel is mainly characterized with the primary driver of growth, besides it is argued for environmental degradation. On the contrary, the receovery in environmental quality stems from renewable energy consumption.[1] Therefore, the associated causality between energy consumption, economic growth, and environmental support researches to comprehend the structural changes in the economy, namely the inflection of inverse U-shaped curve. Furthermore, one of the most important reasons behind the presence of the urbanization or any other social indicators in EKC equation is their power to indicate the structural shifts of developing countries.

METHODOLOGY

The fundamental spectral decomposition techniques of both financial and economic time series are Fourier and wavelet transformations. The Fourier transform, one of the spectral decomposition techniques, is a transformation technique that allows the analysis of the content of a signal in the time domain. In Fourier transform, a function in the time domain is transformed into a function in the frequency domain. Then the Fourier coefficients of the function are analyzed for each

---

[1]Ozturk (2010), Omri (2014) and Tiba and Omri (2017) provide detailed literature reviews for the studies that investigate the relationship between economic growth, energy consumption and environmental degradation.

frequency (Graps, 1995). Basically, sinusoidal waves for the Fourier series are chosen as the main function. The properties of the emerging dilatation are examined in Burrus, Gopinath, and Guo (1998). The mathematical expression of the Fourier transform is explained by the following equations:

In a function f (£) with period T (Illing, 2008):

$$f(£ + T) = f(£) \tag{10.1}$$

To make the function *2pi* periodic, we can redefine the value of £. A new independent variable, then, is defined as, $t = \frac{2pi}{T}£$, as a result:

$$f(t + 2pi) = f(t) \tag{10.2}$$

Since the function is periodic, we only need to consider behavior in a range of length *2pi*, for example on the interval $(-pi, \ pi)$. After Joseph Fourier (1768−1830), the series of Fourier has been transformed into Eq. (10.3), taking into account the infinite sum of sine and cosine functions.

$$f(x) = \frac{a_0}{2} + \sum_{m=1}^{\infty} [a_m \cos(mt) + b_m \sin(mt)] \tag{10.3}$$

In Eq. (10.4), constant coefficients of a_m and b_m are called Fourier coefficients. Estimation with integers using the orthogonality properties of a_m and b_m coefficients of Eq. (10.3) are presented in Eqs. (10.4) and (10.5).

$$a_m = \frac{1}{pi} \int_{-pi}^{pi} f(t)\cos(mt)dx \tag{10.4}$$

$$b_m = \frac{1}{pi} \int_{-pi}^{pi} f(t)\sin(mt) \, dx \tag{10.5}$$

As it can be seen from the equations above, the Fourier series is a linear component of sinuses and cosines. Each of these sinuses and cosines is a function of frequency. Therefore, the Fourier transform can be considered a separation on the basis of frequency by frequency (Gençay, Selçuk, & Whitcher, 2002).

On the other hand, Fourier analysis can localize signal in frequency domain successfully as it cannot use signal in time domain. It gives the knowledge of the whole time span and does not reveal the time-point information at a particular point. Fourier transformation is, therefore, a successful approach to the analysis of time-invariant signals. That is, the Fourier basic function (sinuses and cosines) is more suitable when working with stationary time series. However, restricting ourselves to stationary time series is not very appealing because most economic/financial time series exhibit highly complex patterns (e.g., complex trends, sudden changes, and volatility clustering). Conventional signal analysis tools, such as Fourier analysis, miss these frequency components. For this reason, in a nonstationary time series, Fourier transformation cannot effectively detect complex events (Gençay et al., 2002). For example, the Fourier transformation is not suitable for the nonstationary series, and the transformation is insufficient because it gives the whole time-domain knowledge, not the information of the specific time unit. Because of this deficiency, the wavelet transform is used for complex events through scale-based analyses.

Wavelet transformation is a more recent and more complicated methodology compared with Fourier transformation. It has similar mathematical representation with Fourier transform; however, it comes up with a new property, called scaling. Wavelet has the advantage of localizing signals both in time and frequency domain simultaneously. The wavelet transform, as the best technique for the nonstationary time series, is filtered into different frequency bands, which are divided into segments in the time domain (Zhao, Jiang, Diao, & Qian, 2004). The wavelet function (or mother wavelet function) can be expressed as a linear combination of the scaling function. Nonetheless, with condition $\omega \in \mathbb{R}$ and $m \in \mathbb{R}^+$, every scaled and translated wavelet function $\beta_{(m,\omega)}(t)$ can be written in terms of the mother wavelet:

$$\beta_{(m,\omega)}(t) = \frac{1}{\sqrt{|m|}}\beta\left(\frac{t-\omega}{m}\right) \qquad (10.6)$$

where the term $1/\sqrt{|m|}$ denotes normalization, factor ensuring unit variance of wavelet. The mother wavelet $\beta(.)$ contains two control parameters m (scaled) and ω (located). The parameter m controls the width of the wavelet and indicates the position of the wavelet. Parameter ω is the translation or location parameter that controls the location of wavelet and represents the position of the wavelet in the time domain. When evaluated in terms of frequency, rapidly changing details capture at lower scales, namely high frequencies, while slower details capture higher scales, namely low frequencies. In other words, an increasing m (scaling) capture low frequency (long-run) properties of time series; however, a descending m compresses it to measure high-frequency (short-run) properties. This situation indicates a negative relationship between the frequency and the scale.

The continuous wavelet transform (CWT) is obtained by projection a specific wavelet $\beta(.)$ onto the examined time series $\ddot{X}(t) \in L^2(\mathbb{R})$ in regard to wavelet $\beta_{(m,\omega)}(t)$ is written as:

$$W_{\ddot{X}}(m,\omega) = \int_{-\infty}^{\infty} \yen(t)\frac{1}{\sqrt{|m|}}\beta^*\left(\frac{t-\omega}{m}\right)dt \qquad (10.7)$$

where $W_{\ddot{X}}(m,\omega)$ represents CWT and $*$ denotes complex conjugation. Because of a function to be accepted as CWT, it must carry out the following conditions (Conraria and Soares, 2014);

- The integral of $\beta(.)$ is zero,

$$\int_{-\infty}^{\infty} \beta(t)dt = 0, \qquad (10.8)$$

- It is square of $\beta(.)$ integrates to unity,

$$\int_{-\infty}^{\infty} |\beta(t)|^2 dt = 1, \quad \text{and,} \qquad (10.9)$$

- Admissibility condition as is given in Eq. (10.10).

$$\yen_\partial = \int_0^{\infty} \frac{|\theta(\mathcal{N})|^2}{\mathcal{N}}d\mathcal{N} < \infty, \qquad (10.10)$$

Wavelets defined as a small wave generally do not have the same properties. In this respect, wavelets are divided into different groups according to their particular features. These types of wavelets include Haar, Daubechies, Symlets, Coiflets, Biorthogonal, Gauss, Mexican Hat, Morlet, Meyer, Shannon, etc. The Morlet wavelet transform has both imaginary and real parts, thereby it allows to analysis both phase and amplitude. The Morlet wavelet, a sine wave damped by a Gaussian envelope, is expressed as:

$$\lambda_\varphi(t) = pi^{-1/4}\left(e^{i\varphi t} - e^{-\varphi^2/2}\right)e^{-t^2/2}, \qquad (10.11)$$

where parameter φ denotes the central frequency parameter of Morlet wavelet $\lambda_\varphi(t)$. Moreover, $e^{i\varphi t}$ is the complex exponential function, and it is normalized by $pi^{-1/4}$ that provides a function with unit energy (Addison, 2002). $e^{-\varphi^2/2}$ is the correction term, which corrects the nonzero mean of the complex sinusoid, and it is negligible if $\varphi > 5$. This yields a simplified version of Morlet wavelet function written as:

$$\lambda_\varphi(t) = pi^{-1/4}e^{i\varphi t}e^{-t/2} \qquad (10.12)$$

The value 6 for φ ensures the admissibility condition, where λ is the Fourier period (Farge, 1992). The admissibility condition of wavelets, introduced in Eq. (10.10), is important comes from in fact it ensures that it is possible to recover $\ddot{x}(t)$ from its wavelet transform. When $\emptyset$ is analytic and $\ddot{x}(t)$ real a reconstruction equation formula written as is in Morlet and Grossman (1984),

$$\ddot{x}(t) = (\Psi_\emptyset)^{-1} \int_0^\infty \left[\int_{-\infty}^\infty W_{\ddot{X}}(m, \omega) \phi_{(m,\omega)}(t) d\omega \right] \frac{dm}{m^2} \quad (10.13)$$

In addition to CWT, preserves the energy or variance of preservation of the examined time series;

$$\|\ddot{x}\|^2 = (\Psi_\emptyset)^{-1} \int_0^\infty \left[\int_{-\infty}^\infty \left| W_{\ddot{X}}(m, \omega) \right|^2 d\omega \right] \frac{dm}{m^2} \quad (10.14)$$

where, the wavelet power spectrum $\left| W_{\ddot{x}}(m, \omega) \right|^2$, which shows the distribution energy of $\ddot{X}(t)$ of time series in both frequency and time domain, is shown. In addition, this feature is used to define the wavelet correlation, wavelet variance, and wavelet covariance. The cross-wavelet power[2] of two time series, $W_{xy}(m, \omega)$ can be defined as the local covariance between these time series at each scale (frequency band) and each time. The cross-wavelet power of two time series $x(t)$ and $y(t)$ was first presented by Hudgins, Friehe, and Mayer (1993) as below:

$$W_{xy}(m, \omega) = W_x(m, \omega) \overline{W_y(m, \omega)} \quad (10.15)$$

where $W_x(m, \omega)$ and $W_y(m, \omega)$ are CWT of time series $x(t)$ and $y(t)$ as $W_{xy}(m, \omega)$ is the cross-wavelet power. In addition, m is scale and ω is location parameter as they appear in CWT formula in Eq. (10.7). After, that is, two time series, the cross-wavelet transforms represent the local covariance between the time series at each scale (Vacha & Barunik, 2012). According to Aguiar-Conraria, Magalhães, and Soares (2013), the wavelet coherency of two time series, $W_x(m, \omega)$ and $W_y(m, \omega)$ can be defined as follows:

$$R_{xy}(m, \omega) = \frac{\left| S\left(W_{xy}(m, \omega) \right) \right|}{\left| S\left(\left| W_{xx}(m, \omega) \right| \right) \right|^{1/2} \left| S\left(\left| W_{yy}(m, \omega) \right| \right) \right|^{1/2}} \quad (10.16)$$

where R_{xy} represents the correlation, parameter ranging from one (strong consistency) to zero coherency (no

---

[2]When $x = y$, we obtain the wavelet power spectrum.

coherency) in both time and frequency domain. In addition, S refers to the required smoothing parameter. Otherwise, coherency will always be strong consistency (one). The phase difference analysis detects phase relationships between components, for instance, the correlation direction (positive and negative correlation), and lead or lag relation. The phase difference (with $\xi_{x,y} \in [-pi, pi]$) between time series $x(t)$ and $y(t)$ can be depicted as:

$$\xi_{x,y}(m, \omega) = \frac{1}{\tan} \left(\frac{\Im\left(W_{xy}(m, \omega) \right)}{\Re\left(W_{xy}(m, \omega) \right)} \right) \quad (10.17)$$

In Eq. (10.17), $\Im(W_{xy})$ and $\Re(W_{xy})$ are referred to imaginary and real parts, respectively. If $\xi_{xy} \in \left(0, \frac{pi}{2} \right)$, the series move in phase and $x(t)$ leads $y(t)$; If $\xi_{xy} \in \left(0, -\frac{pi}{2} \right)$, the series move again in phase, then, $y(t)$ is leading. If $\xi_{x,y} \in \left(\frac{pi}{2}, pi \right)$, there is antiphase relation, in this case, the series move again out of phase where $y(t)$ is leading. Antiphase relation exists, when phase difference is pi or $-pi$. If $\xi_{x,y} \in \left(-pi, -\frac{pi}{2} \right)$, then, the series follow antiphase relation as $x(t)$ is leading. Finally, a phase difference of zero indicates that $y(t)$ and $x(t)$ move together.

DATA AND WAVELET ESTIMATION OUTPUT

This work considers mainly the inspecting the comovements between GDI and CO_2 (million metric tons of CO_2) in a wavelet model for the United States for the quarterly period from 1980:1 to 2018:2.

The wavelet model in this work first monitors the influence of GDI on CO_2, later, uses as well control variables to be able to capture cleaner output. The control variables are total fossil fuels consumption (fossil, quadrillion Btu), nuclear electric power consumption (nuclear, quadrillion Btu), and total renewable energy consumption (renewables, quadrillion Btu), respectively, for the United States for the same period.

The CO_2, fossil, nuclear, and renewables have been extracted from US Energy Information Administration Monthly Energy Review (EIA, 2018), and the GDI data have been obtained from Fred data through ESTIMA-RATS 9.1 (Federal Reserve Bank of St. Louis, 2018 August). All estimations have been launched by MATLAB 15 program lines.

Fig. 10.1 depicts the US GDI for the period 1980: 1−2018:2, and Fig. 10.2 exhibits the US CO_2 emissions (million metric tons) for the period 1980: 1−2018:2. Fig. 10.1 explores that GDI of United States tends to increase from the first quarter of 1980 and the second quarter of 2018. Through polynomial trend

Gross Domestic Income (GDI), 1980:1-2018:2

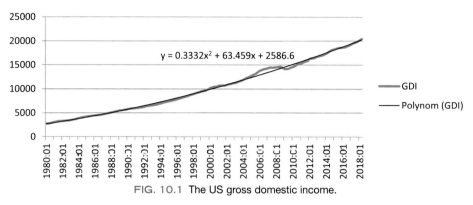

FIG. 10.1 The US gross domestic income.

Carbon Dioxide (CO_2) emissions (million metric tons), 1980:1-2018:2

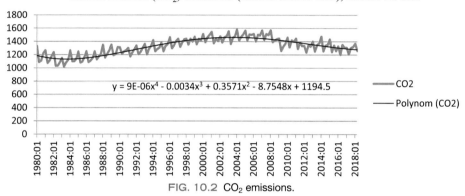

FIG. 10.2 CO_2 emissions.

estimation, one may state that GDI increases by $[63.459 + 0.3332(2x + 1)]$. CO_2 emissions firsts decline until 1984, later increases till 2016, and afterward tends to reduce. The estimated polynomial regression of CO_2 is $y = 9E{-}06x^4 - 0.0034x^3 + 0.3571x^2 - 8.7548x + 1194.5$.

Figs. 10.3−10.5 denote the trends of the fossil fuel consumption (quadrillion Btu), nuclear electric power consumption (quadrillion Btu), and, renewable energy consumption (quadrillion Btu), respectively, in the United States for the period 1980:1−2018:2.

The fossil fuel consumption, nuclear electric power consumption, and renewable energy consumption follow the fourth degree polynomial, third degree of polynomial, and fourth degree of polynomial, respectively. Fossil fuel consumption tends to increase until 2006 and afterward diminishes. Consumption of renewables tends to enlarge at increasing rate, as nuclear electric consumption expands at decreasing rate from March of 1980 to June of 2018.

Estimating the continuous wavelet models, this section reveals the movements between the US GDI and US CO_2 emissions through time series and frequency analyses. Throughout wavelet estimations, one might inspect the trends of CO_2 as GDI increases at different time periods and frequencies.

Figs. 10.6A, 10.7A, and 10.8A show the outputs of wavelet coherency analyses. Figs. 10.6B and C, 10.7B and C, and, 10.8B and C explore the phase differences, which exhibit the radians between two waves at the same frequency and same time point.

All wavelet analyses in this work aim at observing what happens to level of CO_2 as GDI increases at different time periods and different time frequencies.

In wavelet estimations, the black curve (contour) exhibits 5% significance level of the estimation through an ARMA (p, q) model. AR (p) and MA (q) terms of the ARMA model depict the autoregressive terms and moving average (lagged forecast errors) terms, respectively. Our estimations used ARMA (1, 1) and ARMA

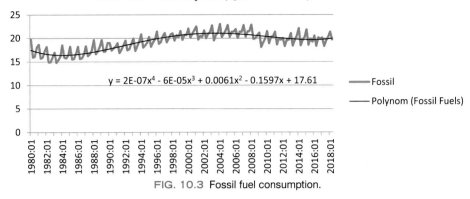

FIG. 10.3 Fossil fuel consumption.

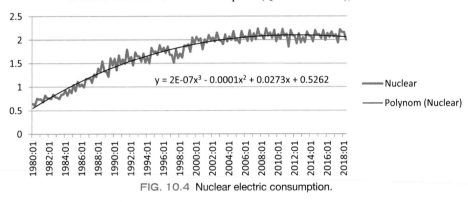

FIG. 10.4 Nuclear electric consumption.

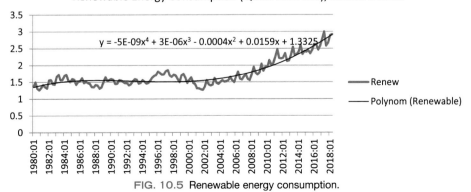

FIG. 10.5 Renewable energy consumption.

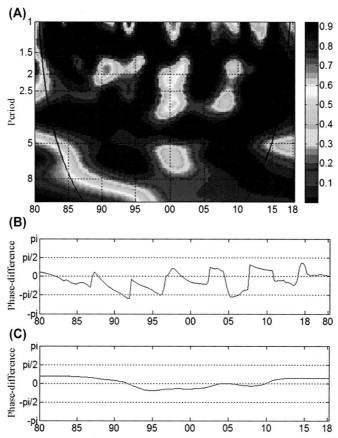

FIG. 10.6 **(A)** Wavelet coherence (GDI-CO₂), 1980:1−2018:2. **(B)** 1−3 frequency band, 1980:1−2018:2. **(C)** 3−8 frequency band, 1980:1−2018:2.

(2, 1) models. The colors to the right of the figures represent the degree of coherency between the variables. It ranges from weak coherency (blue) to strong coherency (red) between the variables. The color code, therefore, explores as well the possible weakest coherence (dark blue) and strongest coherence (dark red). The power of correlation ranging from 0.05 to 0.95 might be considered energy of association. The dark red and dark blue, thereby, correspond to high energy of association and low energy of association between the variables, respectively.

Fig. 10.6A reveals the wavelet coherency between (GDI-CO₂) for the period 1980:1−2018:2. Fig. 10.6A and its phase difference (Fig. 10.6B) reveal little evidence of significant effect of GDI on CO₂ during the first half of the 1990s at 1−3 year frequency band. Figs. 10.6A and C denote that there exists significant association between GDI and CO₂ emissions in 8-year cycle during the second half of the 1980s and the first

half of the 1990s and that GDI enhances CO₂ emissions during the same frequency and time points.

When relevant control variables have been added to wavelet estimations, the coherency analyses have become clearer. Figs. 10.7A and 10.8A observe the movements of relevant pairs of variables by considering the relevant control variables of fossil fuels consumption (Fig. 10.7A) and total energy consumption (Fig. 10.8A). Total energy consumption variable includes the consumptions of fossil fuels, nuclear, and renewables.

Fig. 10.7A indicates that (a) GDI and CO₂ have strong energy association during the first half of the 1990s and 2010s in 1-year cycle (frequency); (b) GDI and CO₂ are slightly associated between years 2003 and 2005 in 1-year cycle; and (c) GDI and CO₂ appear to have strong coherency during the first half of the 1980s, the second half of the 1990s, and, for the period 2012−16 at 3-year frequency.

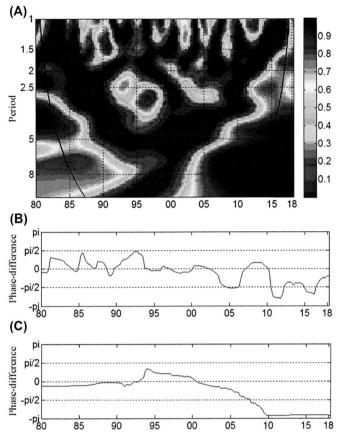

FIG. 10.7 **(A)** Wavelet coherence (GDI-CO_2 || fossil), 1980:1–2018:2. **(B)** 1–3 frequency band, 1980:1–2018: 2. **(C)** 3–8 frequency band, 1980:1–2018:2.

The phase difference (Fig. 10.7B) demonstrates that, as GDI increases, the CO_2 increases during the period 1990–94, later, diminishes in 2011–12, and in 2015–16.

As for longer time period (3–8 year frequency), GDI and CO_2 variables present strong coherencies during 1982–90 and 2004–16 (Fig. 10.7C). CO_2 leads GDI during the periods 1982–90 and 2004–07, while GDI causes CO_2 to reduce for the period 2008–16 (Fig. 10.7C).

In Fig. 10.8A, again the wavelet analyses are conducted for the variables of GDI and CO_2 by adding the control variables of fossil, nuclear, and renewable energy consumptions. Considering 1–3 frequency band, strong correlations between GDI and CO_2 are shown in the years 1982–85, 1987–90, 2005, and, 2011–13. The correlation emerges slightly strong during 1998–2000.

Fig. 10.8B yields the evidence that GDI leads CO_2 to increase in 1982–85 and to diminish in 2015. On the other hand, CO_2 is leading positively the GDI variable during periods 1987–90 and 2011–13.

Monitoring 3–8 frequency band in Fig. 10.8A, one might state that GDI and CO_2 follow strong coherencies in time periods of 1982–86 and 2006–16. Fig. 10.8C displays the output that, as GDI increases, CO_2 increases for the period 1982–86 and 2004–06, and shrinks for the period 2011–16.

Following continuous partial wavelet coherency analyses presented in Figs. 10.7A and 10.8A and associated phase difference pictures, we might argue that as GDI of the United States improves initially in the beginning of sample period (second half of 1980 and first half of the 1990s), CO_2 emissions tend to advance, and CO_2 emissions tend to reduce during the quarters of 2008–16.

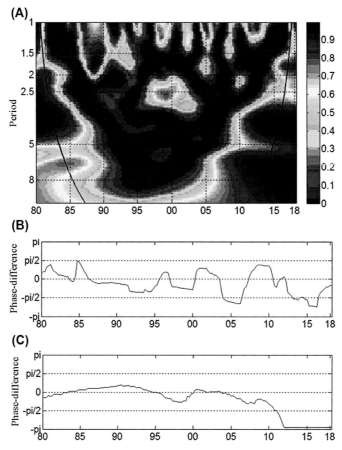

FIG. 10.8 **(A)** Wavelet coherence (GDI-CO_2 ‖ fossil, nuclear, renewable), 1980:1–2018:2. **(B)** 1–3 frequency band, 1980:1–2018:2. **(C)** 3–8 frequency band, 1980:1–2018:2.

CONCLUSION AND POLICY PROPOSALS

The contributions of existing seminal works to the literature of EKC hypotheses are significant. However, time series and panel data estimations of EKC in general follow the same estimated parameters within whole sample period. Although some prominent works consider the structural breaks in cross-section dependence tests of panel data, they reveal eventually constant estimates in observing the effect of GDP (GDI) on environmental degradation for the whole predicted time period. Few articles aim at detecting the relevant estimates of coefficients with one or two structural breaks in a dynamic structure. The dynamic structure of an EKC model can be evaluated through either the effects of leads and lags of independent variable (GDI) on dependent variable (environmental degradation or CO_2 emissions), or, through possible changes in estimated parameters in one or more potential

structural breaks. Furthermore, dynamic structure of an EKC model can be examined through all possible shifts in estimated values at all possible different time periods and at all relevant time frequencies. Such analyses, thereby, observe numerous structural breaks within different time cycles (frequencies) and, hence, might result in more efficient, consistent, and unbiased estimators.

This work first reviews intensively the relevant literature evidence to exhibit the EKC estimations' results through time series and panel data estimations.

The work later, underlining the mathematical advantages of wavelet transformation against Fourier transformation, follows continuous wavelet and partial continuous wavelet coherency analyses.

Fourier analysis can localize signal in frequency domain successfully as it cannot use signal in time domain. Wavelet, on the other hand, has the advantage of localizing signals both in time and frequency domain simultaneously.

All wavelet analyses here aim at observing what happens to level of CO_2 emissions as GDI increases at different time periods and at different time frequencies in the United States from the first quarter of 1980 to the second quarter of 2018.

In conclusion, following the quarterly time period 1980:1−2018:2, this book reveals that, as GDI of the United States increases, initially in the beginning of sample period (second half of 1980 and first half of 1990s), CO_2 emissions tend to increase, and, CO_2 emissions tend to decline during the quarters of 2008−16.

This output confirms the EKC hypothesis indicating that the level of environmental deterioration gets worse and later becomes better in the United States, as domestic income of the United States goes up.

Increase in income of representative countries might improve the societies' awareness of environmental deterioration within regions/countries. This statement might be confirmed due to new technologic innovations, more efficient usage of natural resources through environmental awareness, improved market mechanism, and, throughout considerations of optimal intertemporal choices of consumption preferences.

The policy-makers and researchers might inspect the role of components of total energy consumption on environmental quality. For instance, in recent literature, some works underline the positive effects of renewables on environmental improvements as depicted in Ulucak and Bilgili (2018), Bilgili, Koçak, Bulut, and Kuloğlu (2017), Bilgili, Koçak, Bulut, and Kuşkaya (2017), Bilgili, Koçak et al. (2016), and Bilgili, Öztürk et al. (2016). They explore the significant positive impact of renewables on quality level of environment in terms of CO_2 emissions. They recommend that renewables, such as consumption of wood and agricultural products, solid waste, landfill gas and biogas, ethanol, and biodiesel should be stimulated to reduce the environmental degradation. By following these works and other seminal researches, this work might suggest that the US policy-makers continue to follow existing and potential future US energy policies.

Indeed, the US energy policies, since the 1970s, have been following the goals of (1) conservation and energy efficiency, (2) efficient domestic supply of fossil fuels, and (3) efficient production/consumption of electricity and electricity from renewables (Ballotpedia, 2018a). The United States launched Environmental Protection Agency (EPA) in 1971, Comprehensive Environmental Response, Compensation, and Liability Act (CERCLA) in 1980, Kyoto Protocol in 1997, and, American Recovery and Reinvestment Act (2009) as explained in Ballotpedia (2018b).

On the one hand, the United States has been implementing energy policies to protect environment since the 1970s; on the other hand, the US population has been increasing its demand for goods and commodities since then. This might, in turn, cause the United States to have increasing ecologic deficit, which is the difference between biocapacity and ecologic footprints.

However, according to wavelet analyses of this work, the United States has succeeded to improve her environmental quality after the second half of 2000s and during 2010s due to past and current energy policies of the United States.

Among available policies, the Energy Policy Acts of the United States might have relatively more considerable positive impacts on environment among other energy policies/acts. Hence, this book eventually recommends that policy-makers consider following and updating 1992 Energy Policy Act (EPACT) and 2005 EPACT to lower the CO_2 emissions to reach sustainable intertemporal and intergenerational development.

APPENDIX

Notes (1): Causality Signs: ⇔ (bidirectional causality); ⇒ (uni-directional causality from left to right); ∼ (mixed results) and//(lack of causality).

Notes (2): Abbreviations for variables: AEC (alternative energy consumption); BEC (biomass consumption); CE (CO_2 emissions); CTAX (carbon tax); EC (energy consumption); ECD (disaggregated energy consumption); ECF (ecological carbon footprint); ED (energy demand); EDEG (environmental degradation); EG (economic growth); EI (energy intensity); ELC (electricity consumption); ENP (energy price); EX (exports); FD (financial development); FDI (foreign direct investment); FFC (fossil fuel consumption); FIXC (fixed capital stock); GLB (globalization level); HEC (hydropower consumption); IGT (intermediate goods trade); IM (imports); IND (industrialization level); INV (investment expenditures); M1 (money supply); NE (nuclear energy); NGC (natural gas consumption); NREC (non-renewable energy consumption); NREP (non-renewable electricity production); PC (petroleum consumption); POP (population density); REC (renewable energy consumption); RELC (renewable energy consumption); REP (renewable electricity production); RLL (ratio of liquid liabilities to national income); TO (trade openness); TOD (tourism development); TRD (international trade), TRL (trade liberalization); TVA (transport sector value added); UR (urbanization); WPD (water pollution discharge).

Notes (3): Methods: IAAA (innovative accounting approach); AMG (causality and augmented mean group estimator); ARDL (Auto autoregressive distributed lag model); VECM (Vector Error Correction Method).

Notes (4): Estimation methods: OLS (ordinary least squares); FGLS (feasible generalized least squares; LAE (least absolute deviations); GMM (generalized method of moments); STAR (structural threshold regression or auto-regression), GLS (generalized least squares).

REFERENCES

Abdallah, K. B., Belloumi, M., & Wolf, D. D. (2013). Indicators for sustainable energy development: A multivariate cointegration and causality analysis from Tunisian road transport sector. *Renewable and Sustainable Energy Reviews, 25*, 34–43. https://doi.org/10.1016/j.rser.2013.03.066.

Abid, M. (2015). The close relationship between informal economic growth and carbon emissions in Tunisia since 1980: The (ir) relevance of structural breaks. *Sustainable Cities and Society, 15*, 11–21. https://doi.org/10.1016/j.scs.2014.11.001.

Acaravci, A., & Akalin, G. (2017). *Environment–economic growth nexus: A comparative analysis of developed and developing countries# 7.*

Addison, P. S. (2002). The illustrated wavelet transform handbook: Introductory theory and application in science. In *Engineering, medicine and finance* (1st ed.). Taylor&Francis Group, CRS Press.

Adu, D. T., & Denkyirah, E. K. (2018). Economic growth and environmental pollution in West Africa: Testing the environmental Kuznets curve hypothesis. *Kasetsart Journal of Social Sciences*. https://doi.org/10.1016/j.kjss.2017.12.008.

Aguiar-Conraria, L., Magalhães, P. C., & Soares, M. J. (2013). The nationalization of electoral cycles in the United States: A wavelet analysis. *Public Choice, 156*(3–4), 387–408. https://doi.org/10.1007/s11127-012-0052-8.

Aguiar-Conraria, L., & Soares, M. J. (2014). The continuous wavelet transform: Moving beyond uni- and bivariate analysis. *Journal of Economic Surveys, 28*(2), 344–375. https://doi.org/10.1111/joes.12012.

Ahmad, N., Du, L., Lu, J., Wang, J., Li, H. Z., & Hashmi, M. Z. (2017). Modelling the CO_2 emissions and economic growth in Croatia: Is there any environmental Kuznets curve? *Energy, 123*, 164–172.

Ahmad, A., Zhao, Y., Shahbaz, M., Bano, S., Zhang, Z., Wang, S., et al. (2016). Carbon emissions, energy consumption and economic growth: An aggregate and disaggregate analysis of the Indian economy. *Energy Policy, 96*, 131–143. https://doi.org/10.1016/j.enpol.2016.05.032.

Ahmed, K. (2017). Revisiting the role of financial development for energy-growth-trade nexus in BRICS economies. *Energy, 128*, 487–495. https://doi.org/10.1016/j.energy.2017.04.055.

Ahmed, K., & Long, W. (2012). Environmental Kuznets curve and Pakistan: An empirical analysis. *Procedia Economics and Finance, 1*, 4–13.

Ahmed, K., Rehman, M. U., & Ozturk, I. (2017). What drives carbon dioxide emissions in the long-run? Evidence from selected South Asian countries. *Renewable and Sustainable Energy Reviews, 70*, 1142–1153. https://doi.org/10.1016/j.rser.2016.12.018.

Ahmed, K., Shahbaz, M., Qasim, A., & Long, W. (2015). The linkages between deforestation, energy and growth for environmental degradation in Pakistan. *Ecological Indicators, 49*, 95–103. https://doi.org/10.1016/j.ecolind.2014.09.040.

Ajmi, A. N., Hammoudeh, S., Nguyen, D. K., & Sato, J. R. (2015). On the relationships between CO_2 emissions, energy consumption and income: The importance of time variation. *Energy Economics, 49*, 629–638. https://doi.org/10.1016/j.eneco.2015.02.007.

Akbostancı, E., Türüt-Aşık, S., & Tunç, G. I. (2009). The relationship between income and environment in Turkey: Is there an environmental Kuznets curve? *Energy Policy, 37*(3), 861–867. https://doi.org/10.1016/j.enpol.2008.09.088.

Al-Mulali, U., & Ozturk, I. (2016). The investigation of environmental Kuznets curve hypothesis in the advanced economies: The role of energy prices. *Renewable and Sustainable Energy Reviews, 54*, 1622–1631. https://doi.org/10.1016/j.rser.2015.10.131.

Al-Mulali, U., Ozturk, I., & Solarin, S. A. (2016). Investigating the environmental Kuznets curve hypothesis in seven regions: The role of renewable energy. *Ecological Indicators, 67*, 267–282. https://doi.org/10.1016/j.ecolind.2016.02.059.

Al-mulali, U., Weng-Wai, C., Sheau-Ting, L., & Mohammed, A. H. (2015). Investigating the environmental Kuznets curve (EKC) hypothesis by utilizing the ecological footprint as an indicator of environmental degradation. *Ecological Indicators, 48*, 315–323. https://doi.org/10.1016/j.ecolind.2014.08.029.

Aldy, J. E. (2005). An environmental Kuznets curve analysis of U.S. State-level carbon dioxide emissions. *The Journal of Environment & Development, 14*(1), 48–72. https://doi.org/10.1177/1070496504273514.

Ali, W., Abdullah, A., & Azam, M. (2017). Re-visiting the environmental Kuznets curve hypothesis for Malaysia: Fresh evidence from ARDL bounds testing approach. *Renewable and Sustainable Energy Reviews, 77*, 990–1000. https://doi.org/10.1016/j.rser.2016.11.236.

Apergis, N. (2016). Environmental Kuznets curves: New evidence on both panel and country-level CO_2 emissions. *Energy Economics, 54*, 263–271. https://doi.org/10.1016/j.eneco.2015.12.007.

Apergis, N., & Ozturk, I. (2015). Testing environmental Kuznets curve hypothesis in Asian countries. *Ecological Indicators, 52*, 16–22. https://doi.org/10.1016/j.ecolind.2014.11.026.

Arouri, M. E., Youssef, A. B., Mhenni, H., & Rault, C. (2012). Energy consumption, economic growth and CO_2 emissions in Middle East and North African countries. *Energy Policy, 45*, 342–349. https://doi.org/10.1016/j.enpol.2012.02.042.

Aslanidis, N., & Iranzo, S. (2009). Environment and development: Is there a Kuznets curve for CO_2 emissions? *Applied Economics*, *41*(6), 803–810. https://doi.org/10.1080/00036840601018994.

Azomahou, T., Laisney, F., & Van, P. N. (2006). Economic development and CO_2 emissions: A nonparametric panel approach. *Journal of Public Economics*, *90*(6–7), 1347–1363. https://doi.org/10.1016/j.jpubeco.2005.09.005.

Baek, J. (2015). A panel cointegration analysis of CO_2 emissions, nuclear energy and income in major nuclear generating countries. *Applied Energy*, *145*, 133–138. https://doi.org/10.1016/j.apenergy.2015.01.074.

Ballotpedia. (2018a). *Energy policy in the United States*. https://ballotpedia.org/Energy_policy_in_the_United_States.

Ballotpedia. (2018b). *Environmental policy in the United States*. https://ballotpedia.org/Environmental_policy_in_the_United_States.

Bello, M. O., Solarin, S. A., & Yen, Y. Y. (2018). The impact of electricity consumption on CO_2 emission, carbon footprint, water footprint and ecological footprint: The role of hydropower in an emerging economy. *Journal of Environmental Management*, *219*, 218–230. https://doi.org/10.1016/j.jenvman.2018.04.101.

Bento, J. P., & Moutinho, V. (2016). CO_2 emissions, non-renewable and renewable electricity production, economic growth, and international trade in Italy. *Renewable and Sustainable Energy Reviews*, *55*, 142–155. https://doi.org/10.1016/j.rser.2015.10.151.

Bertinelli, L., & Strobl, E. (2005). The environmental Kuznets curve semi-parametrically revisited. *Economics Letters*, *88*(3), 350–357. https://doi.org/10.1016/j.econlet.2005.03.004.

Bhattarai, M., & Hammig, M. (2004). Governance, economic policy, and the environmental Kuznets curve for natural tropical forests. *Environment and Development Economics*, *9*(3), 367–382. https://doi.org/10.1017/s1355770x03001293.

Bilgili, F., Koçak, E., & Bulut, Ü. (2016). The dynamic impact of renewable energy consumption on CO_2 emissions: A revisited environmental Kuznets curve approach. *Renewable and Sustainable Energy Reviews*, *54*, 838–845. https://doi.org/10.1016/j.rser.2015.10.080.

Bilgili, F., Koçak, E., Bulut, Ü., & Kuloğlu, A. (2017). The impact of urbanization on energy intensity: Panel data evidence considering cross-sectional dependence and heterogeneity. *Energy*, *133*, 242–256. https://doi.org/10.1016/J.ENERGY.2017.05.121.

Bilgili, F., Koçak, E., Bulut, Ü., & Kuşkaya, S. (2017). Can biomass energy be an efficient policy tool for sustainable development? *Renewable and Sustainable Energy Reviews*, *71*, 830–845. https://doi.org/10.1016/j.rser.2016.12.109.

Bilgili, F., Öztürk, İ., Koçak, E., Bulut, Ü., Pamuk, Y., Muğaloğlu, E., et al. (2016). The influence of biomass energy consumption on CO_2 emissions: A wavelet coherence approach. *Environmental Science and Pollution Research*, *23–19*, 19043–19061. https://doi.org/10.1007/s11356-016-7094-2.

Bo, S. (2011). A literature survey on environmental Kuznets curve. In , *Energy Procedia: Vol. 5. 2010 international conference on energy, environment and development – ICEED 2010* (pp. 1322–1325). https://doi.org/10.1016/j.egypro.2011.03.229.

Boden, T., Andres, R., & Marland, G. (2017). *Global, regional, and national fossil-fuel CO_2 emissions (1751–2014) (V. 2017)*. https://doi.org/10.3334/CDIAC/00001_V2017.

Borhan, H., & Ahmed, E. M. (2012). Green environment: Assessment of income and water pollution in Malaysia. *Procedia – Social and Behavioral Sciences*, *42*, 166–174. https://doi.org/10.1016/j.sbspro.2012.04.178.

Borhan, H., Ahmed, E. M., & Hitam, M. (2012). The impact of CO_2 on economic growth in Asean 8. *Procedia – Social and Behavioral Sciences*, *35*, 389–397. https://doi.org/10.1016/j.sbspro.2012.02.103.

Bo, L., Suying, F., Junbiao, Z., & Haishan, Y. (2011). Carbon functions of agricultural land use and economy across China: A correlation analysis. *Energy Procedia*, *5*, 1949–1956. https://doi.org/10.1016/j.egypro.2011.03.336.

Boutabba, M. A., Diaw, D., & Lessoua, A. (2018). Environment-energy-growth nexus in Sub-Saharan Africa: The role of intermediate goods. *International Economics*. https://doi.org/10.1016/j.inteco.2018.04.003.

Bruyn, S. D., Bergh, J. V., & Opschoor, J. (1998). Economic growth and emissions: Reconsidering the empirical basis of environmental Kuznets curves. *Ecological Economics*, *25*(2), 161–175. https://doi.org/10.1016/s0921-8009(97)00178-x.

Burrus, C. S., Gopinath, R. A., & Guo, H. (1998). *Introduction to wavelets and wavelet transforms a primer*. New Jersey: Prentice Hall.

Charfeddine, L. (2017). The impact of energy consumption and economic development on ecological footprint and CO_2 emissions: Evidence from a Markov switching equilibrium correction model. *Energy Economics*, *65*, 355–374. https://doi.org/10.1016/j.eneco.2017.05.009.

Charfeddine, L., & Mrabet, Z. (2017). The impact of economic development and social-political factors on ecological footprint: A panel data analysis for 15 MENA countries. *Renewable and Sustainable Energy Reviews*, *76*, 138–154. https://doi.org/10.1016/j.rser.2017.03.031.

Cole, M. A. (2004). Trade, the pollution haven hypothesis and the environmental Kuznets curve: Examining the linkages. *Ecological Economics*, *48*(1), 71–81. https://doi.org/10.1016/j.ecolecon.2003.09.007.

Coondoo, D., & Dinda, S. (2002). Causality between income and emission: A country group-specific econometric analysis. *Ecological Economics*, *40*, 351–367. https://doi.org/10.1016/S0921-8009(01)00280-4.

Culas, R. J. (2007). Deforestation and the environmental Kuznets curve: An institutional perspective. *Ecological Economics*, *61*(2–3), 429–437. https://doi.org/10.1016/j.ecolecon.2006.03.014.

Danish, B., Zhang, B., Wang, B., & Wang, Z. (2017). Role of renewable energy and non-renewable energy consumption on EKC: Evidence from Pakistan. *Journal of Cleaner Production*, *156*, 855–864. https://doi.org/10.1016/j.jclepro.2017.03.203.

Dijkgraaf, E., & Vollebergh, H. R. J. (2001). *A note on testing for environmental Kuznets curves. OCFEB research memorandum 0103, 'environmental policy, economic reform and endogenous technology'*. Working paper series 7.

Dinda, S. (2004). Environmental Kuznets curve hypothesis: A survey. *Ecological Economics, 49,* 431−455. https://doi.org/10.1016/j.ecolecon.2004.02.011.

Dinda, S., Coondoo, D., & Pal, M. (2000). Air quality and economic growth: An empirical study. *Ecological Economics, 34*(3), 409−423. https://doi.org/10.1016/s0921-8009(00)00179-8.

Dogan, E., & Seker, F. (2016). The influence of real output, renewable and non-renewable energy, trade and financial development on carbon emissions in the top renewable energy countries. *Renewable and Sustainable Energy Reviews, 60,* 1074−1085. https://doi.org/10.1016/j.rser.2016.02.006.

Dong, K., Sun, R., & Hochman, G. (2017). Do natural gas and renewable energy consumption lead to less CO_2 emission? Empirical evidence from a panel of BRICS countries. *Energy, 141,* 1466−1478. https://doi.org/10.1016/j.energy.2017.11.092.

EIA, & US Energy Information Administration. (2018). *Total energy data.* https://www.eia.gov/totalenergy/data/monthly/.

Ersin, Ö.Ö. (2016). The nonlinear relationship of environmental degradation and income for the 1870-2011 period in selected developed countries: The dynamic panel-STAR approach. *Procedia Economics and Finance, 38,* 318−339. https://doi.org/10.1016/s2212-5671(16)30205-2.

Ertugrul, H. M., Cetin, M., Seker, F., & Dogan, E. (2016). The impact of trade openness on global carbon dioxide emissions: Evidence from the top ten emitters among developing countries. *Ecological Indicators, 67,* 543−555. https://doi.org/10.1016/j.ecolind.2016.03.027.

Esteve, V., & Tamarit, C. (2012). Threshold cointegration and nonlinear adjustment between CO_2 and income: The Environmental Kuznets Curve in Spain, 1857−2007. *Energy Economics, 34*(6), 2148−2156. https://doi.org/10.1016/j.eneco.2012.03.001.

Farge, M. (1992). Wavelet transforms and their applications to turbulence. *Annual Reviews, 24,* 395−457. https://doi.org/10.1146/annurev.fl.24.010192.002143.

Farhani, S., & Shahbaz, M. (2014). What role of renewable and non-renewable electricity consumption and output is needed to initially mitigate CO_2 emissions in MENA region? *Renewable and Sustainable Energy Reviews, 40,* 80−90. https://doi.org/10.1016/j.rser.2014.07.170.

Federal Reserve Bank of St Louis. (2018). *Federal Reserve Bank of St. Louis | economic data, monetary rates.* https://www.stlouisfed.org/.

Galeotti, M., & Lanza, A. (1999). Richer and cleaner? A study on carbon dioxide emissions in developing countries. *Energy Policy, 27*(10), 565−573. https://doi.org/10.1016/s0301-4215(99)00047-6.

Galeotti, M., & Lanza, A. (2005). Desperately seeking environmental Kuznets. *Environmental Modelling & Software, 20*(11), 1379−1388. https://doi.org/10.1016/j.envsoft.2004.09.018.

Gao, B. (2011). The impacts of economic growth on resources and environment in henan province. *Procedia Environmental Sciences, 11,* 810−816. https://doi.org/10.1016/j.proenv.2011.12.124.

Gençay, R., Selçuk, F., & Whitcher, B. (2002). *An introduction to wavelets and other filtering methods in finance and economics.* San Diego, CA: Academic Press.

Graps, A. (1995). An Introduction to wavelets. *IEEE Computational Science and Engineering, 2,* 50−51. https://doi.org/10.1109/99.388960.

Grossman, G., & Krueger, A. (1991). *Environmental impacts of a North American free trade agreement.* https://doi.org/10.3386/w3914.

Grossman, G. M., & Krueger, A. B. (1995). Economic growth and the environment. *Quarterly Journal of Economics, 110,* 353−377 https://doi.org/10.2307/2118443.

Halkos, G. E. (2003). Environmental Kuznets Curve for sulfur: Evidence using GMM estimation and random coefficient panel data models. *Environment and Development Economics, 8*(04), 581−601. https://doi.org/10.1017/s1355770x0300317.

Hamit-Haggar, M. (2012). Greenhouse gas emissions, energy consumption and economic growth: A panel cointegration analysis from Canadian industrial sector perspective. *Energy Economics, 34*(1), 358−364. https://doi.org/10.1016/j.eneco.2011.06.005.

Harbaugh, W. T., Levinson, A., & Wilson, D. M. (2002). Reexamining the empirical evidence for an environmental Kuznets curve. *The Review of Economics and Statistics, 84*(3), 541−551. https://doi.org/10.1162/003465302320259538.

He, J., & Richard, P. (2010). Environmental Kuznets curve for CO_2 in Canada. *Ecological Economics, 69*(5), 1083−1093. https://doi.org/10.1016/j.ecolecon.2009.11.030.

Hettige, H., Mani, M., & Wheeler, D. (2000). Industrial pollution in economic development: The environmental Kuznets curve revisited. *Journal of Development Economics, 62*(2), 445−476. https://doi.org/10.1016/s0304-3878(00)00092-4.

Hitam, M. B., & Borhan, H. B. (2012). FDI, growth and the environment: Impact on quality of life in Malaysia. *Procedia − Social and Behavioral Sciences, 50,* 333−342. https://doi.org/10.1016/j.sbspro.2012.08.038.

Holtz-Eakin, D., & Selden, T. M. (1995). Stoking the fires? CO_2 emissions and economic growth. *Journal of Public Economics, 57*(1), 85−101. https://doi.org/10.1016/0047-2727(94)01449-x.

Hudgins, L., Friehe, C. A., & Mayer, M. E. (1993). Wavelet transform and atmospheric turbulence. *Physical Review Letters, 71*(20), 3279−3282. https://doi.org/10.1103/PhysRevLett.71.3279.

Illing, L. (2008). *Fourier analysis.* http://www.reed.edu/physics/courses/Physics33 1.f08/pdf/Fourier.pdf.

Iwata, H., Okada, K., & Samreth, S. (2011). A note on the environmental Kuznets curve for CO_2: A pooled mean group approach. *Applied Energy, 88*(5), 1986−1996. https://doi.org/10.1016/j.apenergy.2010.11.005.

Jalil, A., & Feridun, M. (2011). The impact of growth, energy and financial development on the environment in China: A cointegration analysis. *Energy Economics, 33*(2), 284−291. https://doi.org/10.1016/j.eneco.2010.10.003.

Jalil, A., & Mahmud, S. F. (2009). Environment Kuznets curve for CO_2 emissions: A cointegration analysis for China. *Energy Policy, 37,* 5167−5172. https://doi.org/10.1016/j.enpol.2009.07.044.

Jayanthakumaran, K., Verma, R., & Liu, Y. (2012). CO_2 emissions, energy consumption, trade and income: A comparative analysis of China and India. *Energy Policy, 42,* 450−460. https://doi.org/10.1016/j.enpol.2011.12.010.

Jebli, M. B., Youssef, S. B., & Ozturk, I. (2016). Testing environmental Kuznets curve hypothesis: The role of renewable and non-renewable energy consumption and trade in OECD countries. *Ecological Indicators, 60,* 824−831. https://doi.org/10.1016/j.ecolind.2015.08.031.

Jin, L., Duan, K., Shi, C., & Ju, X. (2017). The impact of technological progress in the energy sector on carbon emissions: An empirical analysis from China. *International Journal of Environmental Research and Public Health, 14,* 1505. https://doi.org/10.3390/ijerph14121505.

Kaika, D., & Zervas, E. (2013). The Environmental Kuznets Curve (EKC) theory—Part A: Concept, causes and the CO_2 emissions case. *Energy Policy, 62,* 1392−1402. https://doi.org/10.1016/j.enpol.2013.07.131.

Katircioğlu, S. T. (2014). Testing the tourism-induced EKC hypothesis: The case of Singapore. *Economic Modelling, 41,* 383−391. https://doi.org/10.1016/j.econmod.2014.05.028.

Kaufmann, R. K., Davidsdottir, B., Garnham, S., & Pauly, P. (1998). The determinants of atmospheric SO2 concentrations: Reconsidering the environmental Kuznets curve. *Ecological Economics, 25*(2), 209−220. https://doi.org/10.1016/s0921-8009(97)00181-x.

Kharbach, M., & Chfadi, T. (2017). CO_2 emissions in Moroccan road transport sector: Divisia, Cointegration, and EKC analyses. *Sustainable Cities and Society, 35,* 396−401. https://doi.org/10.1016/j.scs.2017.08.016.

Koop, G., & Tole, L. (1999). Is there an environmental Kuznets curve for deforestation? *Journal of Development Economics, 58*(1), 231−244. https://doi.org/10.1016/s0304-3878(98)00110-2.

Kunnas, J., & Myllyntaus, T. (2007). The environmental Kuznets curve hypothesis and air pollution in Finland. *Scandinavian Economic History Review, 55*(2), 101−127. https://doi.org/10.1080/03585520701435970.

Kuznets, S. (1955). Economic growth and income inequality. *The American Economic Review, 45,* 1−28.

Lee, C., Chiu, Y., & Sun, C. (2009). Does one size fit all? A reexamination of the environmental Kuznets curve using the dynamic panel data approach. *Review of Agricultural Economics, 31*(4), 751−778. https://doi.org/10.1111/j.1467-9353.2009.01465.x.

Le, T., & Quah, E. (2018). Income level and the emissions, energy, and growth nexus: Evidence from Asia and the Pacific. *International Economics.* https://doi.org/10.1016/j.inteco.2018.03.002.

Lindmark, M. (2002). An EKC-pattern in historical perspective: Carbon dioxide emissions, technology, fuel prices and growth in Sweden 1870−1997. *Ecological Economics, 42*(1−2), 333−347. https://doi.org/10.1016/s0921-8009(02)00108-8.

List, J. A., & Gallet, C. A. (1999). The environmental Kuznets curve: Does one size fit all? *Ecological Economics, 31*(3), 409−423. https://doi.org/10.1016/s0921-8009(99)00064-6.

Liu, Z., Guan, D., Crawford-Brown, D., Zhang, Q., He, K., & Liu, J. (2013). Energy policy: A low-carbon road map for China [WWW Document] *Nature.* https://doi.org/10.1038/500143a.

Loganathan, N., Shahbaz, M., & Taha, R. (2014). The link between green taxation and economic growth on CO_2 emissions: Fresh evidence from Malaysia. *Renewable and Sustainable Energy Reviews, 38,* 1083−1091. https://doi.org/10.1016/j.rser.2014.07.057.

Managi, S., & Jena, P. R. (2008). Environmental productivity and Kuznets curve in India. *Ecological Economics, 65*(2), 432−440. https://doi.org/10.1016/j.ecolecon.2007.07.011.

Martínez-Zarzoso, I., & Bengochea-Morancho, A. (2004). Pooled mean group estimation of an environmental Kuznets curve for CO_2. *Economics Letters, 82*(1), 121−126. https://doi.org/10.1016/j.econlet.2003.07.008.

Morlet, J., & Grossmann, A. (1984). Decomposition of hardy functions into square integrable wavelets of constant shape. *SIAM Journal on Mathematical Analysis, 15*(4), 723−736. https://doi.org/10.1137/0515056.

Moutinho, V., & Robaina, M. (2016). Is the share of renewable energy sources determining the CO_2 kWh and income relation in electricity generation? *Renewable and Sustainable Energy Reviews, 65,* 902−914. https://doi.org/10.1016/j.rser.2016.07.007.

Nahman, A., & Antrobus, G. (2005). The environmental Kuznets curve: A literature survey. *South African Journal of Economics, 73,* 105−120. https://doi.org/10.1111/j.1813-6982.2005.00008.x.

Nasir, M., & Rehman, F. U. (2011). Environmental Kuznets Curve for carbon emissions in Pakistan: An empirical investigation. *Energy Policy, 39*(3), 1857−1864. https://doi.org/10.1016/j.enpol.2011.01.025.

Omri, A. (2014). An international literature survey on energy-economic growth nexus: Evidence from country-specific studies. *Renewable and Sustainable Energy Reviews, 38,* 951−959. https://doi.org/10.1016/j.rser.2014.07.084.

Omri, A., Daly, S., Rault, C., & Chaibi, A. (2015). Financial development, environmental quality, trade and economic growth: What causes what in MENA countries. *Energy Economics, 48,* 242−252. https://doi.org/10.1016/j.eneco.2015.01.008.

Onafowora, O. A., & Owoye, O. (2014). Bounds testing approach to analysis of the environment Kuznets curve hypothesis. *Energy Economics, 44,* 47−62. https://doi.org/10.1016/j.eneco.2014.03.025.

Ozcan, B. (2013). The nexus between carbon emissions, energy consumption and economic growth in Middle East countries: A panel data analysis. *Energy Policy, 62,* 1138−1147. https://doi.org/10.1016/j.enpol.2013.07.016.

Ozokcu, S., & Özdemir, Ö. (2017). Economic growth, energy, and environmental Kuznets curve. *Renewable and Sustainable Energy Reviews, 72,* 639−647. https://doi.org/10.1016/j.rser.2017.01.059.

Ozturk, I. (2010). A literature survey on energy−growth nexus. *Energy Policy, 38,* 340−349. https://doi.org/10.1016/j.enpol.2009.09.024.

Panayotou, T. (1993). *Empirical tests and policy analysis of environmental degradation at different stages of economic development*. ILO Work. Pap. 45. http://sci-hub.tw/; http://www.ilo.org/public/libdoc/ilo/1993/93B09_31_engl.pdf.

Panayotou, T., Peterson, A., & Sachs, J. (2000). *Is the environmental Kuznets curve driven by structural change? What extended time series may imply for developing countries?*. CAER II Discussion Paper No. 80.

Pao, H., & Tsai, C. (2011). Multivariate Granger causality between CO_2 emissions, energy consumption, FDI (foreign direct investment) and GDP (gross domestic product): Evidence from a panel of BRIC (Brazil, Russian Federation, India, and China) countries. *Energy, 36*(1), 685–693. https://doi.org/10.1016/j.cnergy.2010.09.041.

Roberts, J., & Grimes, P. E. (1997). Carbon intensity and economic development 1962–1991: A brief exploration of the environmental Kuznets curve. *World Development, 25*(2), 191–198. https://doi.org/10.1016/s0305-750x(96)00104-0.

Saboori, B., & Sulaiman, J. (2013a). CO_2 emissions, energy consumption and economic growth in association of Southeast Asian Nations (ASEAN) countries: A cointegration approach. *Energy, 55*, 813–822. https://doi.org/10.1016/j.energy.2013.04.038.

Saboori, B., & Sulaiman, J. (2013b). Environmental degradation, economic growth and energy consumption: Evidence of the environmental Kuznets curve in Malaysia. *Energy Policy, 60*, 892–905. https://doi.org/10.1016/j.enpol.2013.05.099.

Saboori, B., Sulaiman, J., & Mohd, S. (2012). Economic growth and CO_2 emissions in Malaysia: A cointegration analysis of the environmental Kuznets curve. *Energy Policy, 51*, 184–191. https://doi.org/10.1016/j.enpol.2012.08.065.

Sahli, I., & Rejeb, J. B. (2015). The environmental Kuznets curve and corruption in the Mena region. *Procedia - Social and Behavioral Sciences, 195*, 1648–1657. https://doi.org/10.1016/j.sbspro.2015.06.231.

Saleem, H., Jiandong, W., Zaman, K., Elashkar, E. E., & Shoukry, A. M. (2018). The impact of air-railways transportation, energy demand, bilateral aid flows, and population density on environmental degradation: Evidence from a panel of next-11 countries. *Transportation Research Part D: Transport and Environment, 62*, 152–168. https://doi.org/10.1016/j.trd.2018.02.016.

Selden, T. m, & Song, D. (1994). Environmental quality and development: Is there a Kuznets curve for air pollution emissions? *Journal of Environmental Economics and Management, 27*, 147–162. https://doi.org/10.1006/jeem.1994.1031.

Seppala, T., Haukioja, T., & Kaivo-oja, J. (2001). The EKC hypothesis does not hold for direct material flows: Environmental Kuznets Curve hypothesis tests for direct material flow in five industrial countries. *Population and Environment, 23*(2), 217–238.

Shafik, N., & Bandyopadhyay, S. (1992). *Economic Growth and Environmental Quality, Background Paper for World Development Report, Working Papers*, (p. 904).

Shafik, N. (1994). Economic development and environmental quality: An econometric analysis. *Oxford Economic Papers, 46*, 757–773.

Shahbaz, M., & Sinha, A. (2016). Environmental Kuznets curve for CO_2 emission: A literature survey. Munich pers. *RePEc Arch*.

Shahbaz, M., Dube, S., Ozturk, I., & Jalil, A. (2015). Testing the environmental Kuznets curve hypothesis in Portugal. *International Journal of Energy Economics and Policy, 5*, 475–481.

Shahbaz, M., Khraief, N., Uddin, G. S., & Ozturk, I. (2014). Environmental Kuznets curve in an open economy: A bounds testing and causality analysis for Tunisia. *Renewable and Sustainable Energy Reviews, 34*, 325–336. https://doi.org/10.1016/j.rser.2014.03.022.

Shahbaz, M., Lean, H. H., & Shabbir, M. S. (2012). Environmental Kuznets curve hypothesis in Pakistan: Cointegration and Granger causality. *Renewable and Sustainable Energy Reviews, 16*(5), 2947–2953. https://doi.org/10.1016/j.rser.2012.02.015.

Shahbaz, M., Ozturk, I., Afza, T., & Ali, A. (2013). Revisiting the environmental Kuznets curve in a global economy. *Renewable and Sustainable Energy Reviews, 25*, 494–502. https://doi.org/10.1016/j.rser.2013.05.021.

Shahbaz, M., Solarin, S. A., Hammoudeh, S., & Shahzad, S. J. (2017). Bounds testing approach to analyzing the environment Kuznets curve hypothesis with structural beaks: The role of biomass energy consumption in the United States. *Energy Economics, 68*, 548–565. https://doi.org/10.1016/j.eneco.2017.10.004.

Shahbaz, M., Solarin, S. A., & Ozturk, I. (2016). Environmental Kuznets Curve hypothesis and the role of globalization in selected African countries. *Ecological Indicators, 67*, 623–636. https://doi.org/10.1016/j.ecolind.2016.03.024.

Shahbaz, M., Solarin, S. A., Sbia, R., & Bibi, S. (2015). Does energy intensity contribute to CO_2 emissions? A trivariate analysis in selected African countries. *Ecological Indicators, 50*, 215–224. https://doi.org/10.1016/j.ecolind.2014.11.007.

Shahbaz, M., Uddin, G. S., Rehman, I. U., & Imran, K. (2014). Industrialization, electricity consumption and CO_2 emissions in Bangladesh. *Renewable and Sustainable Energy Reviews, 31*, 575–586. https://doi.org/10.1016/j.rser.2013.12.028.

Shahzad, S. J., Kumar, R. R., Zakaria, M., & Hurr, M. (2017). Carbon emission, energy consumption, trade openness and financial development in Pakistan: A revisit. *Renewable and Sustainable Energy Reviews, 70*, 185–192. https://doi.org/10.1016/j.rser.2016.11.042.

Shi, J. (2004). *Test of the EKC hypothesis using CO_2 panel data*. Resource Economics and Policy Analysis (REPA). Research Group Working Paper 2004-03.

Solarin, S. A., Al-Mulali, U., & Ozturk, I. (2017). Validating the environmental Kuznets curve hypothesis in India and China: The role of hydroelectricity consumption. *Renewable and Sustainable Energy Reviews, 80*, 1578–1587. https://doi.org/10.1016/j.rser.2017.07.028.

Stern, D. I. (2004). The rise and fall of the environmental Kuznets curve. *World Development, 32*, 1419–1439. https://doi.org/10.1016/j.worlddev.2004.03.004.

Stern, D. I., & Common, M. S. (2001). Is there an environmental Kuznets curve for sulfur? *Journal of Environmental Economics and Management, 41*(2), 162–178. https://doi.org/10.1006/jeem.2000.1132.

Stern, D. I., Common, M. S., & Barbier, E. B. (1996). Economic growth and environmental degradation: The environmental Kuznets curve and sustainable development. *World Development, 24*, 1151–1160. https://doi.org/10.1016/0305-750X(96)00032-0.

Suri, V., & Chapman, D. (1998). Economic growth, trade and energy: Implications for the environmental Kuznets curve. *Ecological Economics, 25*(2), 195–208. https://doi.org/10.1016/s0921-8009(97)00180-8.

Tang, C. F., & Tan, B. W. (2015). The impact of energy consumption, income and foreign direct investment on carbon dioxide emissions in Vietnam. *Energy, 79*, 447–454. https://doi.org/10.1016/j.energy.2014.11.033.

Tan, F., Lean, H. H., & Khan, H. (2014). Growth and environmental quality in Singapore: Is there any trade-off? *Ecological Indicators, 47*, 149–155. https://doi.org/10.1016/j.ecolind.2014.04.035.

Tiba, S., & Omri, A. (2017). Literature survey on the relationships between energy, environment and economic growth. *Renewable and Sustainable Energy Reviews, 69*, 1129–1146. https://doi.org/10.1016/j.rser.2016.09.113.

Tiwari, A. K., Shahbaz, M., & Hye, Q. M. (2013). The environmental Kuznets curve and the role of coal consumption in India: Cointegration and causality analysis in an open economy. *Renewable and Sustainable Energy Reviews, 18*, 519–527. https://doi.org/10.1016/j.rser.2012.10.031.

Torras, M., & Boyce, J. K. (1998). Income, inequality, and pollution: A reassessment of the environmental Kuznets curve. *Ecological Economics, 25*(2), 147–160. https://doi.org/10.1016/s0921-8009(97)00177-8.

Ulucak, R., & Bilgili, F. (2018). A reinvestigation of EKC model by ecological footprint measurement for high, middle and low income countries. *Journal of Cleaner Production, 188*, 144–157. https://doi.org/10.1016/j.jclepro.2018.03.191.

Vacha, L., & Barunik, J. (2012). Co-movement of energy commodities revisited: Evidence from wavelet coherence analysis. *Energy Economics, 34*(1), 241–247. https://doi.org/10.1016/j.eneco.2011.10.007.

Xu, J. (2011). Is there causality from investment for real estate to carbon emission in China: A cointegration empirical study. *Procedia Environmental Sciences, 5*, 96–104. https://doi.org/10.1016/j.proenv.2011.03.054.

Youssef, A. B., Hammoudeh, S., & Omri, A. (2016). Simultaneity modeling analysis of the environmental Kuznets curve hypothesis. *Energy Economics, 60*, 266–274. https://doi.org/10.1016/j.eneco.2016.10.005.

Zambrano-Monserrate, M. A., Silva-Zambrano, C. A., Davalos-Penafiel, J. L., Zambrano-Monserrate, A., & Ruano, M. A. (2018). Testing environmental Kuznets curve hypothesis in Peru: The role of renewable electricity, petroleum and dry natural gas. *Renewable and Sustainable Energy Reviews, 82*, 4170–4178. https://doi.org/10.1016/j.rser.2017.11.005.

Zhang, C., Wang, Y., Song, X., Kubota, J., He, Y., Tojo, J., et al. (2017). An integrated specification for the nexus of water pollution and economic growth in China: Panel cointegration, long-run causality and environmental Kuznets curve. *The Science of the Total Environment, 609*, 319–328. https://doi.org/10.1016/j.scitotenv.2017.07.107.

Zhao, G., Jiang, D., Diao, J., & Qian, L. (2004). Application of wavelet time-frequency analysis on fault diagnosis for steam turbine. In *5th international conference of acoustical and vibratory surveillance methods and diagnostic techniques*. Senlis, France: CETIM. https://doi.org/10.1006/mssp.2002.1517.

Zilio, M., & Recalde, M. (2011). GDP and environment pressure: The role of energy in Latin America and the Caribbean. *Energy Policy, 39*(12), 7941–7949. https://doi.org/10.1016/j.enpol.2011.09.049.

Zoundi, Z. (2017). CO_2 emissions, renewable energy and the Environmental Kuznets Curve, a panel cointegration approach. *Renewable and Sustainable Energy Reviews, 72*, 1067–1075. https://doi.org/10.1016/j.rser.2016.10.018.

CHAPTER 11

Conclusion

ILHAN ÖZTÜRK • BURCU ÖZCAN

The environmental Kuznets curve (EKC) is one of the crucial energy-environmental economics research fields, which has been attracted by numerous academics and researchers for the last three decades. EKC-related studies started with Simon Kuznets (1955) who examined the relationship between income inequality and economic growth. Then it continued with Grossman and Krueger (1991) who analyzed the linkage between economic growth and environmental pollution. This book is a unique and valuable guide for students and EKC practitioners. First, the book aims at analyzing environmental pollution problems in the perspective of economics discipline and proposing solutions against those problems. Second, EKC studies are a versatile tool for the investigation of the energy-growth nexus and finance-energy nexus within various frameworks. Last, the book has a systematic order in respect of context, that is, each section is followed by another section. Thus, readers can easily follow and understand the content of the book.

Section 1 provides a brief historic perspective for the EKC hypothesis. Section 2 classifies and explains literature such as early studies, recent studies, and single versus multiple country studies, whereas Section 3 explains the econometrics of the EKC. Separate chapters are devoted to the early studies and more recent studies, single-country and multiple-country analyses with the purpose of making studies in the field as comparable as possible. Because the book provides material for all stages of the EKC analysis, it will be an indispensable tool for both newcomers and experienced researchers. The book is also a valuable tool for other relevant strands of research such as the "energy-growth-environment nexus" and "finance-energy nexus." The book contains 10 chapters that are summarized as follows:

Chapter 1 provides a historic perspective on global environmental concerns raising over time. The "Limits to Growth" Report, the Brundtland Report, the sustainable development notion, and EKC are the essential notions emphasized in the chapter. In the current scientific field, there exists a central research question about whether economic growth is a foe or a friend for the environment. In this regard, analysts or scientists have been trying to answer this question by testing the EKC hypothesis for different countries. In this chapter, focusing on the development and pollution nexus, authors provide a historic perspective, starting from the 18th century and continuing until the 21st century, on the environmental challenges and the struggles undertaken by countries. As well as offering a historic perspective on environmental problems, this chapter also clarifies the reasons behind the inverted U-shaped EKC by explaining the scale, composition, and technique effects.

Chapter 2 analyzes the earlier empirical studies within the EKC literature between 1991 and 2006. The chapter highlights the major findings from the existing literature, indicating the limitations of these studies. In conclusion, some factors are accepted as the key reasons behind the EKC such as income elasticity of environmental quality demand, scale effects, technologic effects, and composition effects. The chapter ends up with policy implications concerning with the EKC hypothesis.

As presented in Chapter 3, the relationship between environmental degradation and economic development is rather complicated. The literature is characterized by a lack of profound conclusion on the existence of such a trend in single- and multicountry studies. For this reason, the inclusion of important factors is imperative in some cases to capture particular environmental and socioeconomic characteristics of studied cases and potentially can affect the relationship between emissions and economic development. This

Environmental Kuznets Curve (EKC). https://doi.org/10.1016/B978-0-12-816797-7.00011-4

141

chapter presents and discusses a few of them such as tourism activities, financial development, and energy consumption. The discussion revolves around the role that these factors play in the traditional bivariate EKC context and summarizes general findings of recent studies.

Chapter 4 presents the empirical studies in the existing literature. For this purpose, these studies are divided into two groups, as the single-country and the multiple-country studies. In addition, all studies are evaluated according to their results. The literature review indicates that the EKC hypothesis has a long history and is still up to date.

As presented in Chapter 5, the irreversible threshold for global warming, $2°C$ rise in average temperature, is being gradually approached. Humanity will face with a dark and dangerous future if required actions are not taken regarding sustainable development. Although the global community has come a long way in combating environmental threats (e.g., global warming, climate change, ozone depletion, biodiversity loss), there is still much to be done. The purpose of this chapter is first to review the international efforts on the process of sustainable development and environmental awareness and second to discuss the findings of empirical studies that focused on alternative investigations to reach efficient and permanent solution for the threats. The chapter concludes with nonenergy recommendations.

Chapter 6 indicates that there is no single or best solution to the challenge of encouraging the utilization of renewable energy. Rather, picking the most effective options will depend on the specific conditions of each region and market, as well as the particular characteristics of each type of renewable energy. There are, however, a number of available policy options that have been tested in various places around the world. That base of experience should prove helpful for policy-makers who wish to expand the use of renewable energy. At the same time, this chapter also emphasizes the discussion about mobilizing private finance for clean energy innovation. The literature has neglected the structure and regulation of financial markets as potential determinants. To mobilize resources to break out of the fossil fuel technology complex and to finance radical and transformative innovations, the industry needs intermediation to take different channels.

Chapter 7 discusses the problems faced by researchers in the course of EKC analysis and have put forth some basic solutions to handle those problems. The problems are categorized under five broad categories, namely (a) model selection, (b) model validation, (c) data standardization, (d) variable selection and creation, and (e) proxy for environmental degradation. After a brief discussion of the problems, authors try to provide some possible solutions to the problems using real data. The problems discussed in this chapter are mostly application oriented, and thus the solutions provided are indicative, as with the course of time and with the more econometric development, researchers will be able to come up with better solutions to these problems.

Chapter 8 highlights the general issues of time-series analysis that are related to test the EKC hypothesis. Specifically, they cover the issue of spurious regression, stationarity, unit root with and without structural breaks, cointegration with and without structural breaks. They have a feel that the researchers who test the EKC hypothesis in the country case should incorporate more recent econometric methodologies.

Chapter 9 summarizes both conventional and contemporary panel data estimation techniques (e.g., unit root tests, cointegration tests, causality tests, and long-run estimators) used in examining the EKC hypothesis. Previous studies on the relationship between pollution and economic growth have reached consistent or conflicting results depending on the utilization of different estimation techniques. The mixed findings are likely due to econometric problems that cannot be fully captured or solved by a single methodology. However, following the development of various advanced methodologies, such as Common Correlated Effects Mean Group (CCEMG) and Augmented Mean Group (AMG), the researchers have been able to provide justifications in explaining and predicting the EKC phenomenon.

Chapter 10 analyzes the EKC hypothesis by using a different and novel methodology named the "wavelet coherence approach." Based on the wavelet model estimation, authors reexamine the EKC hypothesis for the United States using quarterly data over the period 1980:1−2018:2. This chapter eventually explores that as Gross Domestic Income (GDI) of the United States increases, CO_2 emissions tend to increase in the beginning of sample period, whereas it tends to decline during the last years of the observed period. Hence, for the United States, EKC hypothesis is verified by wavelet coherence estimations considering possible strong and weak associations between GDI and CO_2.

As a conclusion, we would like to thank all contributors of the book who are established and well-known scholars in the EKC research field. In addition, we would like to thank Elsevier team for their help, trust, and constructive suggestions that helped us very much to complete the book. Finally, we believe that this book can help established researchers to realize the new challenges that the EKC field is confronted with.

After following guidelines contained in the book, we believe that researchers in the EKC field will have a great chance to publish their papers in quality journals. Besides, this book can also help readers with no previous experience in the field.

Index

Note: Page numbers followed by "f" indicate figures, "t" indicate tables and "b" indicate boxes.